GENDER AND STEM

This book studies the prevalence of gender inequalities in the domains of science, technology, engineering, and mathematics (STEM) in India. It explores the paradox of an increasing number of women pursuing higher education in STEM fields while a relatively low percentage enter the job market. The book analyzes the factors that contribute to the loss of potential contributions from women in science. It underscores the need to understand the construction of masculinity and femininity, as well as male privilege and female disadvantage and how these concepts are associated with STEM.

Lucid, topical, and rich in empirical data, this book will be indispensable for scholars and researchers of women's studies, gender studies, sociology of science, science and technology studies, political studies, and public policy. It will also be useful to policymakers, organizations, and professionals working in the STEM fields.

Namrata Gupta has been working on gender in science and engineering since 2002 when she obtained her PhD in Sociology (IIT Kanpur). Her research focuses on how the participation and rise of women scientists in academia and research laboratories are affected by the organizational aspects, patriarchal socio-cultural norms, and a lack of universalism in science. Her research also includes the study of women engineering students, the relationship between gender and technology, and gender and high-tech entrepreneurship. She has several publications to her credit in collaboration with renowned international researchers. She has been a recipient of several fellowships and grants from the Indian Council of Social Science Research (ICSSR). She has also been a member of the Editorial Board of the journal *Work, Employment and Society* since 2021 and an Editorial Board member of the UNESCO publication *A Braided River: The Universe of Indian Women in Science* (Jan 2022); and invited to participate in national and international workshops including those funded by the NSF (the US).

GENDER AND STEM

The Indian Context

Namrata Gupta

Routledge
Taylor & Francis Group
LONDON AND NEW YORK

Designed cover image: Getty Images

First published 2025
by Routledge
4 Park Square, Milton Park, Abingdon, Oxon OX14 4RN

and by Routledge
605 Third Avenue, New York, NY 10158

Routledge is an imprint of the Taylor & Francis Group, an informa business

British Library Cataloguing-in-Publication Data
A catalogue record for this book is available from the British Library

ISBN: 978-1-032-97982-3 (hbk)
ISBN: 978-1-032-98021-8 (pbk)
ISBN: 978-1-003-59666-0 (ebk)

DOI: 10.4324/9781003596660

Typeset in Sabon
by Deanta Global Publishing Services, Chennai, India

In memory of my father Shri Rajendra Prasad Singhal

CONTENTS

TABLES

PREFACE

This work is a substantial revision of my earlier book entitled *Women in Science and Technology: Confronting Inequalities*. The reasons for revising the earlier version were compelling. The data and facts have changed since the first edition (2019), which created room for new interpretations and challenging the old ones. The nation went through tough times during the Covid pandemic, including the lockdowns, which affected women's work. Technology is witnessing a rapid pace of growth, and women in engineering and their participation in new technologies needed to be accorded greater attention than in the first edition. New insights on 'gender and STEM' have emerged with the publication of research after the first edition. Also, there has been a substantial revision in the government discourse and policies challenging gender inequality in scientific research, and the new programs of the government, such as GATI, created scope to amend my narrative.

While there is considerable enhancement in programs and policies, gender continues to create barriers for women in STEM education and careers. This book analyzes why the progress remains sluggish through a study of 'social construction' of gender in STEM, with a focus on institutions, such as family, colleges, and STEM workplaces including academia and research institutes. It is my hope that this book will be able to unravel the change and continuity and bring about greater clarity on gender in STEM education and careers.

I thank the team of Taylor and Francis for giving me an opportunity to create this edition and for their support throughout the process.

My eternal thanks to my husband, Vinay, for being a pillar of support; and my daughters, Neha and Manya, for their constant encouragement. The blessings of my parents, and the inspiration that I continue to draw from my (late) father have made this academic journey possible.

ABBREVIATIONS

AIIMS:	All India Institute of Medical Sciences
AISHE:	All India Survey of Higher Education
BPO/ITeS:	Business Process Outsourcing-Information Technology Enabled Services
BTech:	Bachelor of Technology
BDS:	Bachelor of Dental Surgery
CE:	Civil Engineering
CSE:	Computer Science and Engineering
CSIR:	Council of Scientific and Industrial Research
DBT:	Department of Biotechnology
DOS:	Department of Space
DRDO:	Defence Research and Development Organization
DST:	Department of Science and Technology
GATI:	Gender Advancement for Transforming Institutions
GDP:	Gross Domestic Product
GOI:	Government of India
HOD:	Head of the Department
IASc:	Indian Academy of Sciences
IIITs:	Indian Institutes of Information Technology
IISc:	Indian Institute of Science
IIT:	Indian Institute of Technology
INR:	Indian Rupee
ISRO:	Indian Space Research Organization
JEE:	Joint Entrance Exam
KIRAN:	Knowledge Involvement in Research Advancement through Nurturing

MTech: Master of Technology
MBBS: Bachelor of Medicine, Bachelor of Surgery
MD: Doctor of Medicine
MDS: Master in Dental Surgery
ME: Mechanical Engineering
NEET: National Eligibility cum Entrance Test
NIAS: National Institute of Advanced Studies
NIRF: National Institutional Ranking Framework
NIT: National Institute of Technology
PG: Postgraduate
SC: Scheduled Castes
ST: Scheduled Tribes
UG: Undergraduate
UGC: University Grants Commission

INTRODUCTION

Focus and purpose

This work aims to highlight the invisible gender inequalities and unconscious biases in the domain of Indian science, technology, engineering, and mathematics (STEM) and to analyze the relationship between the Indian context and gender inequality that leads to a loss of potential contributions of women in STEM. It suggests that by becoming aware of unconscious biases, the path opens up to tackle those biases.

This book argues that gender is constructed in STEM education and careers, specific to the national cultural context. This is reflected in subject choices, in selection of colleges, and at STEM workplaces, including academia. There is a variation in participation in different STEM fields, and this is often a function of a gendered perception of those fields and the level of economic growth. This is exemplified, for instance, in a high preference among women for computer engineering, compared to other branches of engineering. Further, a high proportion of women in science at the beginning of the pipeline fails to translate into a similar proportion in research. Thus, while the proportion of doctorates awarded to women is about 29% in engineering and technology and 46% in the sciences, there are only about 18.6% of women employed as full-time equivalents in R&D establishments as of April 1, 2021; and of such women, only about one-fourth are in the higher education sector (AISHE 2021-2022: Table 36; S&T Indicators 2023: Table 22). The socio-cultural context is reproduced through formal and informal practices at the workplaces employing science and technology (S&T) personnel, as well as at the academic institutes. The intersectionality of caste and gender complicates the situation of the women of the lowest strata in education and careers.

DOI: 10.4324/9781003596660-1

However, gender is not merely reproduced in the socio-cultural and institutional context; women also actively challenge the context. Further, there is a growing consciousness about harnessing the potential of women in STEM. This is reflected in the multitude of government policies and the change in perspective from enhancing participation to creating gender equality. The coming together of government policy, academia, and women's increasing participation in STEM has created new opportunities for women's deep-tech entrepreneurship, which is a growing area within STEM.

These issues, and a host of others, are explored in this book. The sections following discuss the book's sociological approach that gender is not something 'given' but 'constructed' spatially and historically in a given socio-cultural context.

The book's approach

There are persistent beliefs regarding innate differences between genders, suggesting that there are biological arguments of male superiority for the dominance of men in science and technology (S&T). Dealing with biological arguments and moving from 'sex' to 'gender' to 'social construction of gender' help us to analyze institutions and structures of power that give rise to gender hierarchy in STEM. Women in STEM are disadvantaged by patriarchal norms, masculinity of the professional culture, and an organizational milieu in many countries of the world . However, national cultural beliefs about gender create national similarities and dissimilarities in the manner in which the gender process operates. The process varies not only over space but also over time. Thus, as society and its patriarchal context evolve, so does the relationship between gender and STEM.

Biological versus sociological explanations

Feminists question the traditional belief that women are unfit/less suited to do science or math than men. Technically, while 'sex' refers to the biological/physical distinction of being male/female, 'gender' is patterned; there are socially produced distinctions between male and female, feminine and masculine (Acker 1992). The biological/physical distinctions have been related to few women in higher positions or as high achievers in mathematics and sciences. For instance, Neo-Social Darwinists believe that biological 'difference' is a product of evolution, in which women seem either to have been left behind (in the development of quantitative and spatial abilities) or to lack innate qualities and dispositions suited to science and mathematics (Cronin 2005). However, recent neuroscience studies find that male/female brain differences are trivial and population-specific (Eliot et al. 2021). Further, experience alters brain structures and functioning, so that causal statements about brain differences and success in math and science are circular (Halpern et al. 2007). Biological differences have a sociological explanation and depend on the situational contexts in which women live (Bem 1993). A wide range of socio-cultural forces, including family, neighborhood, peers, and school influence training and experience, and cultural practices contribute to sex differences in mathematics and science achievement and ability (Halpern et al. 2007). Even if men have a higher aptitude than women in mathematics, it would not explain why more women on an average have a higher aptitude than they have careers that require mathematics (Etzkowitz and Gupta 2006). The decision to choose a career in STEM is influenced by the wider context of society and by stereotypes (Powell, Dainty, and Bagilhole 2012). Gender beliefs can bias individuals' expectations regarding their own competence, independently of their underlying abilities. For instance, in analyzing why so few women compared to men choose to pursue highly rewarded STEM careers, Correll (2001) showed that girls attributed less math ability to themselves than did boys with the same math test scores and grades.

The biological arguments do not explain how more than 41% of the total enrolled in doctoral programs in STEM are women (AISHE Report 2021-2022: Figure 19b), but the share of women employed as researchers in STEM is a mere 18%. In a national survey, average mathematics scores for girls and boys of Class X do not show any statistically significant difference (NAS 2018). Also, various scholars working on gender and STEM in India disagree that there is a linking of biological or psychological traits to women's participation in mathematical sciences in the Indian society. These researchers point out that women's innate ability to perform in mathematics is not doubted in India (Mukhopadhyay 1994; Subrahmanyan 1998; Venkatesh 2015).

Nevertheless, there are evidences to the contrary, pointing to the belief in difference in ability based on one's sex. Biological arguments, such as women are more hardworking and that men have higher cognitive skills, are common while attributing a better performance in mathematics (Das and Singhal 2023); these arguments also result in women being discouraged by their professors from applying for higher studies or a female faculty applicant being stringently questioned at interview committees or pushed towards experimental areas of study (Subramanian 2007). Further, gender stereotypes are formed in childhood itself as Kawale and Maji (2022) found in a study of school-going children. The children in the study drew scientist as a 'male' figure, identified scientist as one who works in closed spaces and named primarily men as role model scientists. Stereotypes impact women in education and at the workplace as highlighted in this book. Significantly, men in STEM do not reject stereotypes about a woman's ability to pursue science and engineering with the same vehemence as women (Gupta 2007).

Further, biological arguments in India are closely related to the gendered expectations of parents; and a gendered discourse in families creates male privilege in the STEM fields, even before embarking on a career. The notion of biological differences aligns significantly with the functional differences (Sur 2009) or the argument that since women are biologically different, their roles should differ. The relationship between biological differences and gendered expectations continues to be intrinsic to the Indian society and impacts women scientists and other women professionals through biases at the workplace and in institutions of science.

In general, the education and careers of women in India have been influenced by a patriarchal culture referred to as 'patrifocality' by Mukhopadhyay and Seymour (1994). 'Patrifocality' refers to the kinship and family structures and ideology that give precedence to men over women and imply subordination of an individual to the family. Unlike the system implied in patriarchy, it is not a monolithic system in which males predominate in all settings and social contexts. Rather, 'patrifocality' is the complex linkage between the hierarchical family structure and male dominance and emphasizes lineage purity through control of marriage and female sexuality. Educational decisions and later marital decisions for the children are family decisions. Features of patrifocality include patrilineal inheritance and patrilocal descent and residence that reinforce the centrality of males (for example, a daughter shifts her residence and allegiance to her husband's family after marriage); gender-differentiated family roles (women's nurturing role and domestic roles versus men's economic roles); an ideology of appropriate female behavior emphasizing chastity, obedience, domesticity, and adaptability. Traditionally, concerns of 'purity' (of lineage) through the protection of women's sexuality led to restrictions on male-female interaction (e.g., 'purdah'), which reinforced women's domestic roles, confining them to the 'private' (or domestic) sphere as opposed to the 'public' (or economic) sphere. The purity of caste was contingent upon purity of women (Chakravarti 2004). This ideology also constricted women's education in professional fields and professional employment (Dube 2001; Mukhopadhyay and Seymour 1994). These restrictions have applied more to the middle class and upper castes than others (Subramanyan 1998; Mukhopadhyay and Seymour1994). In effect, the upper castes are more conservative in following the patrifocal ideology (Mukhopadhyay and Seymour 1994).

While there are regional variations in constructions of gender with the North being the patriarchal heartland and the South and the Eastern regions being more diluted in the potency of patriarchy (Datta 2005), there is a broad similarity of patriarchal norms (Mukhopadhyay and Seymour 1994).

Patriarchy is intrinsically linked to women's education that has consequences for women in STEM. Patriarchy itself has not been static over time. To understand this linkage between patriarchy, education, and women in STEM, we first look at the historical context of women's education.

Historical construction of patriarchy and relationship with women's education

In India, the Vedic period (until about 500 B.C.E.) is identified with a period of limited surplus economy coinciding with a lack of deep-seated class inequalities and the caste system in its nascent phase (Desai and Krishnaraj 2004). This early caste system was actually the 'Varna' system in which the Brahmins, Kshatriyas, Vaishyas, and Shudras were hierarchically arranged, with Brahmins at the highest position and the Shudras at the lowest. A lack of stratification in this period is also reflected in the limited nature of gender hierarchy. For instance, the birth of a daughter was not a source of anxiety; women had access to education, which led to the emergence of women scholars such as Lilavati, a mathematician, Gargi, who held philosophical discussions, and of women composers of Vedic hymns (Basham 1967: 179; Basu 1999: 135).

The rigidity of the caste hierarchy in the post-Vedic period is intrinsically related to a decline in the position of women (Desai and Krishnaraj 2004). Gender inequalities acquired normative structure through Manusmriti or the Law Code written around A.D. 200 (Thapar 1966: 121), which codified laws of social structure given by Manu, the ancient lawgiver. The Code established Brahmanical supremacy and declared the oldest male member of the kin as the patriarchal head; women and the Untouchables (positioned outside the Varna system but within the Hindu fold; also referred to as Dalits and officially as Scheduled Castes[1]) were declared unfit for learning. A woman was to be subservient to men, that is, father-husband-son, at different stages of her life (Desai and Krishnaraj 2004). Thus, a decline in women's education went hand-in-hand with the crystallization of a patriarchal system that glorified the birth of a son and elevated the husband to the status of god to be worshipped, along with a host of accompanying rituals that subordinated women. Women's access to education declined steeply over centuries with some exceptions in the Kshatriya and Brahmin families. By the time the British conducted indigenous education surveys in the 1820s and 1830s, there were no girls in traditional village schools or in the schools of higher learning (Basu 1999).

Christian missionaries in the colonial period (A.D. 1800 onwards until 1947) were the first to open schools for girls, followed by the 'liberal' official and non-official English men and women. Wood's Despatch of 1854 was the first official expression of interest in girls' education. Soon, Indian social reformers and women from upper castes/classes, with some notable exceptions, such as Jotiba Phule, also took up the cause of girls' education (Basu 1999). The general view, including those of the Indian social reformers, was that an education of women would aid in their becoming better wives and mothers (Chanana 2001). There were animated debates even among the Indian social reformers about the objectives of educating women. Very few (mostly radicals) argued for women's education from a gender equity point of view, that is, education would overhaul the unjust and hierarchical society. For a majority, women were the custodians of traditional Indian values and their education would help in fulfilling those traditional roles better (Mazumdar 2012: xviii). Separate schools for girls were common in the areas dominated by 'purdah' (Chanana 2001). Many intellectuals of the time favored single-sex schools, for instance, Siqueira (1939: 146) says: 'Experience rather seems to show that boys become girlish and girls boyish when they are educated in mixed schools'. Education for women was seen as an ideal method to smooth over the complexities of the transition from tradition to modernity, to provide an educated wife for men of ambition (Forbes 2008: 69).

Most intellectuals of the time saw gender role differentiation as the basis of curriculum differences, particularly at the secondary education level, justifying women's exclusion from mathematics and science and of men from fine arts (Chanana 2001). It was argued that since women were not expected to take up jobs after education, they should be taught subjects relevant for their domestic roles, such as botany, sociology, and psychology; while, subjects like physics, chemistry, and mathematics should have a subordinate role (e.g., Chiplunkar 1930: 232–241; Siqueira 1939: 142); that secondary education for women should include knowledge of elementary sciences such as botany, zoology, physics, and chemistry 'as applied to household needs and requirements' (Chiplunkar 1930: 236). The ideal science was home science and this was part of the curriculum of the girls' high schools by the 1920s. All India Women's Conference (AIWC), founded in 1927, perceived women's education as important for women's equality, and women were to be educated to be not only better wives but also to be lawyers, doctors, etc. However, this education was to be within the dominant reformist and nationalist framework of the times. Their efforts led to the establishment of Lady Irwin College for Women in 1932 in Delhi, with a certificate course in home science (Basu 2005: 195). A few women such as Menon and Choksi were radicals who believed that education should lead to intellectual development, that physics and mathematics should form part of the curriculum for women (Chanana 2001: 112). Choksi (1929: 68), for example, wrote on the need to alter the curriculum debate because, 'Its [education] purpose should finally be to produce accurate, far-reaching and critical thought'.

Gendered notions about education were held by both colonial officials as well as the Indian intelligentsia. This is best expressed in medical education for women. The medical education for Indian women arose as part of the colonial discourse of 'white man's burden' and control over *zenana* or women's quarters (Burton 2009; Forbes 2009). Much hesitation marked the awarding of medical degrees in the 1870s–1880s to Indian women due to gendered notions about women's intellectual capabilities and their aptitude for science among medical men at the helm of affairs, both Indian and European (Forbes 2009: 6). Kadambini Basu was the first woman to graduate from Calcutta Medical College in 1886. Apart from the bachelor's degree program (MB), there was the Lady Certificate Program requiring the passing of a university entrance examination. Given the social prejudices related to the treatment of Indian women by male doctors, it was deemed imperative that women be treated by women medical practitioners. However, there were not enough women doctors from India or England to staff the hospitals and dispensaries that were established under the Dufferin Fund set up for this purpose in the 1880s (Forbes 2009: 5). Hence, the Hospital Assistant Course was introduced and entry to it required a simple entrance exam with no formal education in science; it was a three-year course which involved little science. Indian intelligentsia did not object to it as women were seen to be different from men, in charge of home and hearth, who needed education suitable to their nature; further, since women's problems were mostly related to childbirth and reproduction, their problems could be dealt with privately by women through this course (Forbes 2009). Thus, the program designed to bring 'Western medicine' to Indian women reflected prejudices related to women's scientific education.

Due to sex-segregation and prejudices against female education, the latter remained mostly an urban phenomenon (Chanana 2001), and all levels of education were heavily male-dominated. There were only 10% of girls of school-going age that were enrolled in schools in British India (Basu 1999). In 1946–1947, there were 36 girls for 100 men in primary schools, 14 in high schools; the numbers tapered off in university education with only seven women for 100 men in colleges of professional education and 12 in colleges of general education and universities (Chanana 2001: 97). Subject-wise, by 1950–1951, the proportion of women to total enrollment in university education was 16.1% in arts, 7.1% in science, 0.16% in technology (19 women enrolled out of total of 12,094), and 16.3% in medicine (Chanana 1993: 129). Thus, arts and medicine had emerged as fields with the maximum participation of women in the period following independence. Women in engineering were unheard of in India at least until the early 1940s. The first women engineer, A Lalitha, graduated in 1943 from the College of Engineering at Guindy.[2]

After independence (1947), all areas of study were thrown open for both women and men. However, secondary education boards of different states continued the exclusion of girls from science and mathematics and of boys from fine arts even in the 1950s. Finally, in 1961, the Committee on Differentiation of Curricula for boys and girls proposed a common curriculum (Mazumdar 2012).

Post-colonial period and the changing patriarchal system

Patriarchy has undergone considerable change since the colonial period. Social practices such as Sati, child marriage and 'purdah' are almost non-existent, and daughters have an equal right to property. Although the 'joint family' continues to be an 'ideal type', the traditional joint family (the most common form of which is the parents, their earning sons and the latters' spouses and children residing together) has declined (Nimkoff 2005). Desai and Andrist (2010) found that marital decisions were being increasingly taken with the consent of the son/daughter in question, although not necessarily based on the bride and groom's familiarity with each other. A step further, there is now a growing trend of women choosing husbands jointly with parents, or a 'hybridization' of Indian and Western practices (Allendorf and Pandian 2016).

However, age and educational hypergamy (groom as more educated, higher in age than bride), the universality of marriage, and a dichotomy of the 'public-private' sphere continue to exist. An ideal woman was a home-maker until a few decades ago. Globalization and the growth of education have brought a greater acceptance of women in public spaces (Thapan 2007). With the growth in education, urbanization, and opportunities for employment, women have greater freedom to work outside the home (Singh 2010). However, the idealization of the perfect woman as primarily devoted to family continues and even received a stimulus with the rise of Hindutva forces in the 1990s, and a pursuit of a highly successful career remains circumscribed for women (Radhakrishnan 2011). This is particularly relevant in the case of middle-class families, which form the focus of our narrative.

The Indian middle class[3] has been typically perceived to be an educated section of urban society who are seeking white-collar jobs or are employed in those types of positions (Sridharan 2004). The educated urban sections, such as the students and professionals in science and engineering, constitute the 'middle class'. For instance, most of the students at prestigious colleges, such as the Indian Institutes of Technology (IITs), are from the middle class as indicated by their father's professions in the JEE-Advanced Report (2018), that collected such statistics for the last time in 2018: fathers of about 44% of students admitted are in the government/private service; about 13% in doctor/engineering/law/teaching, and research/architecture/pharmacy professions; about 8% in agriculture; 15% in business; 3.5% are self-employed; and the rest are 'others'/unknown (JEE Report 2018: 14/3). Although those from income (father's) groups of eight lakhs and above formed only 12% of those registered for the JEE (Advanced), they constituted about 23% of those who attained admission (pg. 14/2). About 77.3% of those admitted belonged to cities, only 12.5% from towns and 10.2% from the villages (pg. 14/1).

Although comprising only about one-third of the population, the Indian middle class has played a conspicuous role ideologically (Fernandes 2006; Deshpande 2003). While multiple identities within this broad 'middle class' exist and are being added, there is a certain 'middleclassness' that defines the attitudes and values of this section due to common patterns of formal education, consumption, and lifestyle aspirations (Donner 2011). For instance, education and careers are extremely important for the middle class as a source of mobility, and jobs are much sought after in the engineering and management fields (Kumar 2011). Sancho (2015) argues that Indian middle-class parents aggressively support their children through their education and build up marketable skills while at the same time appearing to be detached and merely spending money.

Another significant aspect of the 'middleclassness' is the expression of Indianness through a focus on family (Donner 2011; Radhakrishnan 2011). This is evident in the studies of urban women professionals that show a simultaneous existence of individualization and collectivism in their lives. Belliappa (2013) found that professional women in urban India continue to define individual subjectivity in collective terms (an individual self as part of the family). They uphold family as a 'core value' of global Indian identity, and this helps enact 'respectable femininity' at the workplace and in their personal lives (Radhakrishnan 2009). The latest national time-use survey shows that women spend more than three times the number of hours a day than men on unpaid domestic work for household members (NSO 2019). The post-colonial societies do not reflect simply a process of growing individualization; it is complicated by the re-appropriation of tradition. For instance, women's relationships with in-laws are a re-versioning of joint families, with the latter assisting in women's work outside the home, suggesting 'relational reflexivity' or an ability to construct self in relationship to others (Belliappa 2013).

It is in this background that the construction of gender at the STEM institutes and workplaces becomes conspicuous, and this book attempts to unravel the operation of gender inequalities.

STEM and intersectionality

The women in STEM is not a homogeneous group. This book attempts to weave together the intersections of caste, class, and gender, although limited in scope due to a lack of data and studies. Women in STEM are usually urban upper castes and middle class. This is indicated not only by those who qualify for the JEE as mentioned above, but also in studies on women scientists (Gupta 2016, 2017). A NIAS-IAS survey (Kurup et al. 2010) found that about 80% of women in science and in research belong to the General Castes (i.e., upper castes or about 25% of the female population[4]), and about

66% of 'women in research' live in cities. Women working in the information technology (IT) sector are middle class (Radhakrishnan 2009), usually from better economic backgrounds, and more qualified than men (Women and IT Scorecard 2018).

The upper caste background of the women in STEM has historical roots. Traditionally, knowledge has been the purview of the Brahmins. The new elite that arose in the British period, the *bhadralok*, was the first to take up modern Western ideas and Western science. This elite was drawn from the Hindu upper castes (Raj 1991). The social and cultural capital enjoyed by the upper castes enabled them to continue their hegemony over science and scientific enterprise in the post-independence period. The advantage of upper castes in terms of social position and education facilitated women from these families to enter science. Research continues to be dominated by upper caste men and women, particularly in the realm of pure sciences (Chadha and Achuthan 2017).

The middle-class composition of women in science has consequences for them owing to the normative evolution of this class. Modern nationalist culture was projected by the middle class as a binary of 'outer/inner', 'material/spiritual', 'world/home'; and India/East was seen as superior to the West in 'spiritual' (Chatterjee 1989). Purity of the middle-class woman along with a gendered notion of middle-class domesticity was central to this project. The binary was also a way of resolving a preference for Western science while being loyal to Indian nationalism. Acutely conscious of the Western superiority in modern science, Indians faced racism as they were denied access to all higher scientific and technical positions (Basu 1991) and had limited access to research in S&T. According to Partha Chatterjee (1989), nationalism sought to reconcile the desire to pursue Western science and their opposition to colonialism. Through the division of the cultural sphere into material and spiritual, science and technology were seen as part of the 'material', while, 'spiritual' referred to the Indian socio-cultural sphere where women became the chief definers of the 'spiritual' sphere. A scientist in the colonial period was the bearer of 'nationalist masculinity' (Mazumdar 2017: 180). Logically, therefore, the category of 'women' could not fit in this ideological positioning of science as 'material' and masculine.

Nevertheless, influenced by the nationalist movement in colonial times, and encouragement of women's education by the social reformers, women scientists took up the Gandhian way of life, a simple ascetic life, wearing *khadi* (Sur 2009); however, they rejected the Gandhian view that since women are biologically different, their functions should differ and that education should aid their primary functions of homemaker. Adoption of a simple life by women scientists at the time transformed a scientific calling into asceticism (an ascetic being deeply revered in Indian tradition), which helped to create acceptance in public spaces (Sur 2009).

In the immediate post-independence period, the Nehruvian idea of Indianness was defined in secular and socialist terms (Radhakrishnan 2011) that were in contrast to the Hindu rightist ideals. In the Nehruvian India, the old 'middle class' was made up of government workers. Liberalization and globalization since the 1990s have led to the emergence of the 'new middle classes' (Fuller and Narasimhan 2007); this, coupled with the growing Hindu nationalist forces, has revitalized the ideas of the role of Hindu piety, moralism with a central role of family. The new middle class does not differ from the old in terms of their composition, but in how they perceive their position in the nation, characterized by often conservative cultural or nationalist values coupled with a conscientious integration into the global political economy (Radhakrishnan 2009). The post-colonial notion of Indian culture remains gendered with an ideological 'public-private' (economic-domestic) dichotomy, assigning a central role to women in the family.

The dominance of the upper-caste, middle-class group in science continues to date. Caste is not only traditional (delineated in classical Hindu texts), it is linked to power politics (formation of caste associations) and, more importantly, to humiliation (Jodhka 2015). Caste and patriarchy are inextricably linked through the concept of 'honor' that emphasizes the woman's chastity (Chakravarti 1993). Upper caste male advances/violence against Dalit women reinforces caste hierarchy and patriarchy (Gorringe 2018). Inter-caste marriages continue to be frowned upon.

Caste discrimination, although outlawed in the Indian Constitution, exists tenaciously in all walks of life, from schools to universities, in subtle forms. The assumptions of neutrality and rationality in science institutions preclude discourse on caste biases. Caste is a case of acquisition of cultural capital depending upon one's location in the caste position. In the seemingly casteless institutions of S&T, caste is visible in surnames, language, and culture, and the biases in science institutions are extremely difficult to pinpoint, leading Kondaiah et al. (2017) to label them as, '"castelessly" casteist' (p. 77).

The caste disadvantage is reflected in the science and engineering education and careers of the Dalits and Adivasis.[5] These groups are severely underrepresented in engineering courses. Although Dalits and Adivasis constitute 16.6 and 8.6% of the Indian population, respectively, (according to the last Census in 2011; and about 19% and 11%, respectively, as per the latest estimates[6]), the Dalits form 10.6% and the Adivasis 3.6% of the enrollments in BTech (AISHE Report 2021-2022: Table 11). Their representation at the elite institutes is dismal, with Dalits constituting 10% and Adivasis only 2% of those in PhD courses in 2020 at the five high-ranked IITs; the proportion being worse in IISc; and, despite reservation in recruitments, more than 90% of professors in such institutes are from privileged castes (Paliwal 2023). There is a growing government insistence on sticking to caste quota in such positions. However, the filling up of such positions requires action at multiple levels of institutions from schools to colleges. In all of this, gender remains hidden and women's severe underrepresentation among the Dalits/ Adivasis is hardly talked about.

Construction of masculinity: gendered discourse and STEM education

Masculinity can be defined as values, experiences, and meanings that are culturally interpreted as masculine and are ascribed to men more than women in a particular cultural context (Alvesson and Billing 2009: 72). Study of masculinity deepens our understanding of the underlying drivers of gender inequality; it helps to focus on men, patriarchal power structures, and highlights that patriarchy also places dictates on men (Edstrom et al. 2014).

The Indian patriarchal system and the prevalence of extreme forms of masculine ideals form the root of gender inequalities and the process of maintaining dominant masculinity reinforces structural inequalities (Deshmukh-Ranadive 2005; UNESCO 2021). A dominant form of masculinity reflecting Indian patriarchal norms is constructed through early socialization at home and school. The process involves gender socialization, that is, learning society's rules and regulations dichotomized as either masculine or feminine and behaving in ways socially accepted as a male or female; it is in one's family initially that boys realize their privileged status and its relation to power (Adams and Coltrane 2005).

Male privilege and dominance are reproduced in Indian families through differential socialization of sons and daughters, and these differences are rooted in the Indian society's son preference (Srivastava 1997). Traditional Hindu society exhibits son preference as a son is required for cultural rites. Girls are to be married (with dowry) and in the process, lost to parents and, therefore, a liability; while a son carries forward the family name, belongs to the family, takes care of the aged parents and is, therefore, an asset. Srivastava (1997) found that urban mothers (65%), urban grandmothers (72%), rural mothers (74%), and rural grandmothers (96%) overwhelmingly preferred a son to a daughter. Restrictions, pressures, and regulations for girls are enormous; but for boys there is little restriction or opposition for their behaviors, actions, and thoughts and are overprotected and overcared for. While girls tend to conform to the rules of the game, boys use flexible rules and change them to suit their convenience (Chowdhury 2008). In a study (Ram et al. 2014) of male and female youth (aged 15–24 years), it was found that they inhabit different social worlds; that both male and female youth commonly faced gender-discriminatory practices within their households; that sons are given preference in education, freedom to roam, to perform household tasks outside home, have greater freedom to spend money or buy clothes, and are more gender conservative than the female youth.

Schools appear as institutionalized mechanisms for construction of masculinity. Analysis of school curriculum shows that traditional meanings of masculine and feminine continue to persist, such as valorizing achievements of men while those of women are normalized through a focus on their domestic activities (Bhog 2002). Science education itself is value-laden. Raveendran and Chunawala (2015) analyzed science curriculum documents and the NCERT Class XII biology textbook; they found that the curriculum/chapters reflect the masculine, positivist discourse of science and supported technologies that pose a risk to a woman's body. Changes in curriculum are in the offing with the mandate for a new curriculum framework released in April 2022 (Mandate Document 2022), as part of New Education Policy 2020, emphasizing development of human values such as gender equality. However, apart from textbooks, gender stereotypes are also perpetuated through a 'hidden curriculum' (Raina 2012) that includes the classroom environment and teacher biases (Rakshit and Sahoo 2020).

Construction of masculinity is closely associated with educational attainments of boys. Due to the patriarchal emphasis on the male as an asset, there is a greater societal focus and parental expectations regarding science and math achievements of boys than of girls. The discourse on whether the math capability of girls is doubted or not is perhaps of less relevance in India, and it is more important to understand how the society views the achievements of men and women. Thus, even among well-off education-oriented families, educational achievement, especially in the scientific fields, is viewed differently for girls than boys (Mukhopadhyay 1994: 108–109). Men's academic achievement is celebrated more than that of women as it will bring long-term economic and social benefits to the family, while women's achievement has little long-term value for the natal family. Further, even if there is parental support, women moving ahead of men in mathematics and science might not be looked upon favorably among male peers and teachers (Gupta and Sharma 2003).

The patriarchal system constructs masculinity in colleges. Bansal (2013) found that young, urban men (college going, 20–29 years old), freed from the physical proximity of their father's patriarchal authority through hostel life, try to indulge in irreverences. Although not performing too well academically, it is considered a trade-off; peer groups emerge as a source of guidance, support, and information, lowering their reliance on parents. On the other hand, young girls desire to be independent but also emphasize their commitment to family (Thapan 2007); they feel conscious of the need to maintain chastity with respect to sexual relations (Diwakar 2016). Indian norms inhibiting interaction among opposite sexes impede the construction of a self-assured masculinity (Iyer 2014). These have consequences for the informal environment in colleges. There is a lack of understanding of how masculinity is constructed in the professional and higher educational domains through day-to-day interactions and practices. This book attempts to show how the construction of masculinity in these domains creates a male advantage and a female disadvantage in STEM.

Masculinity could also have a negative impact on men. This is particularly due to the gendered expectation of a male as a breadwinner. For instance, boys might be forced to take up the science stream in school due to its job potential (Gautam 2015). Migration abroad for higher studies might be more difficult for men than for women due to the patriarchal expectation of caring for parents (Sondhi 2015). Not all men might subscribe to dominant masculinity, as for instance, in engineering institutes a dominant sexist culture might be disliked by some male students (Gupta 2020). Further, not all men are powerful and those lacking power might be at a receiving end of the dominant masculine culture . However, there is 'the pervasive association of the social category of men with power' (Hearn 2004: 51). This creates a disadvantage for women, particularly in STEM research, due to the association of scientific research with a 'male' model and a masculine culture, as shown in this book.

What this book does

This book analyzes how gender discrimination in the Indian society operates, beginning from the gendered choices at the school stage, to 'doing gender' (West and Zimmerman 1987) in organizations and institutes, with a focus on the formal and informal practices, which reproduce the socio-cultural context. This book also examines the contradictions of a growing number of women in higher education in the fields of STEM but a low percentage in the job sector; and an increasing participation of middle-class, educated, urban women at the workplace but a miniscule number in leadership positions. The book is not a male versus female narrative; rather, it deals with the consequences of patriarchy and the typical Indian socio-cultural context. It highlights the inequalities that are often invisible and aims to make men and women equal partners in contributing to science and technology, according to their full potential.

Conceptually, it adopts the social constructionist perspective. Unlike the feminist viewpoints that consider either men and women as similar (as in the liberal feminist view) or different (as in the radical view), this perspective holds that while there are biological differences between men and women, it is gender or the socially constructed norms, roles, and behaviors that create male-female distinction and defines masculinity or femininity in any society (West and Zimmerman 1987). The cultural male-female difference belief system informs not only interactions, behaviors, and performances, but also institutions and structures, resulting in divergent outcomes for men and women (Ridgeway and Correll 2004). This work shows that men's numeric predominance and a masculine culture at educational institutes and workplaces inhibit women's growth in STEM. Methodologically, the book is substantially based on my own research on women in STEM and also brings together other studies and reports in this field. Most of my work is based on a qualitative methodology to capture the perceptions of women and men in STEM so as to understand how gender is 'done' in institutes/organizations.

The book is organized as follows:

Chapter 1 gives an insight into the higher education system, choice of courses, institutes, and the relationship between gender and educational decisions. It examines how women's choice of science stream in higher secondary school, 'coaching', and later selection of educational institute are a product of a gendered discourse in families. Socio-cultural norms, including those related to masculinity/femininity, tend to be reproduced and reconstructed through such choices/decisions. This is indicated in women's participation numbers that are inversely proportional to the prestige of a course and an institute. While the popularity of computer engineering is growing, the rise of female enrollment in undergraduation in this course is merely 25% compared to a 105% rise among men between 2014–2015 and 2021–2022. However, there are positive trends too, such as the rising proportions of women at the doctoral level in science, engineering, and medical sciences, indicating the growing value attached to women's professional education.

Chapter 2 analyzes the nature of gender biases in engineering education, construction of gender in engineering institutes, and the relationship between gender and caste that create hierarchy among women. Within engineering there are various domains that vary in popularity. Families and recruiters tend to regard certain branches, such as computer engineering, as women-friendly and civil engineering as masculine. The nature of the institute, elite versus non-elite, affects women more than men. A lack of a quality engineering education in the non-IITs/NITs and the process of placements have gendered implications. Further, female students in engineering institutes face performance pressure and other consequences of minority status, such as a lack of informal interaction, social capital, and a voice in student bodies. The female engineering students of the lowest caste have the highest feeling of 'alienation' among all caste groups, and men and women.

Chapter 3 moves on to the STEM professions. The labor market for educated women professionals exhibits gender discrimination in various ways. These include a gender pay gap, bias in hiring and promotions, and few women in leadership positions. This chapter discusses how gender becomes salient at the STEM workplace, which is not only male-dominated but also has a masculine culture. Although women hold a high proportion of degrees in many STEM fields, there are few employed in those professions, as for instance, in architecture and medicine. The proportion of women in the IT workforce appears stagnant since about 2010s. The chapter analyzes how gender is constructed in the major STEM professions of engineering, IT, medicine, and tech entrepreneurship. Issues related to female competence, denial of opportunities due to assumptions of domestic roles, and low access to male-dominated networks are some of the barriers to women's careers common to the STEM professions. Nevertheless, does female participation provide a challenge to the traditional patriarchal norms?

Chapter 4 focuses on inequalities of representation in scientific research and the issues related to productivity and evaluation. Research productivity and citations constitute important criteria of merit. However, is the position of women in science a function of research productivity? Further, it examines the gendered impact of both the international status of Indian research and the local Indian construct of 'merit'. Indian science is practiced in institutes that vary steeply in resources and prestige. How does this hierarchy impact women in science? The second part of the chapter emphasizes that science institutions and individual careers are embedded in the national socio-cultural milieu. Gender interacts with this milieu, creating barriers for women in science in the doctoral period of study, particularly for the women of the lowest caste, for whom the struggle is much more arduous.

Chapter 5 explores how gender is constructed in research careers at the workplace. Although there is no discrimination against women in the written rules related to recruitment or promotions, there are few women in leadership positions and fewer who receive awards. This is because the institutional system of science (as discussed in the previous chapter) and the norms and practices at the workplace fail to create 'merit' for women. At the workplace, problems in informal socialization, the concept of an 'ideal' scientist, and the perception of women as 'privileged', construct male privilege. Women are at a disadvantage due to the colleagues' assumptions of dual burden and a lack of gender-sensitivity in the system. The male advantage, the female disadvantage, and the institutional system of science are the 'triple stressors' for women scientists, which have consequences for their careers. Often, incidents of overt discrimination at the workplace are individualized and gender inequality is seen as 'social', unrelated to the system, which leads to self-blame and precludes discussion on gender inequality at the workplace.

Finally, Chapter 6 analyzes whether gender inequality in science is being questioned and in what ways. Women scientists are not only impacted by gendered norms, structures, and practices; they also challenge gender inequality. This chapter examines the engagement of women scientists with gender inequality issues, how discourse on gender and STEM has shifted in recent times, and with it, the nature of government initiatives. The chapter also suggests the changes in structures and practices that could help in challenging gender inequality. It delineates the changes required in the spheres related to networking, work-life balance, dealing with biological differences, and casteism. It indicates that the best practices elsewhere (e.g. US, Australia, UK) and studies in certain research areas might suggest measures for dealing with gender issues. The chapter also points out the significance of the inclusion of men in the discourse on gender equality and that of the changes in gendered norms in society, which although slow, aid in moving towards greater gender equality in STEM.

Notes

1 Dalits: They belong to the lowest stratum of the Indian society, traditionally excluded from the four-fold Varna system and regarded as untouchables. They are referred to as Scheduled Castes (SC) in the Indian Constitution. Officially, there are three levels of castes for the purpose of reservations: General Castes (or the upper castes), Other Backward Class (OBCs), and the Scheduled Castes and Tribes (the lowest strata). Seats are reserved for the OBCs (the economically disadvantaged layer among them) and the SC/ST in the government institutes.
2 Please see Shantha Mohan's article 'The First Woman Engineer in India', published online on the website of Society of Women Engineers on May 24, 2017. Available at: https://alltogether.swe.org/2017/05/first-indian-woman-engineer/
3 The Indian middle class usually includes an Indian with income ranging from Rs 5 Lakhs – 30 Lakhs p.a. or USD 6,700–40,000. (PRICE 2022: https://www.ice360.in/app/uploads/2022/11/middle-class-brochure-2022.pdf)
4 In the absence of a Census survey since 2011, and based on the latest estimates indicated by the caste composition of primary school students, the General Caste proportion was roughly 25% of the population in 2021–2022. Hence, upper castes' females are 25% of the female population. Please see 'Enrolment by Social Category', Report ID 4004, UDISE+ Reports, Available at: https://dashboard.udiseplus.gov.in/#/reportDashboard/sReport
5 Adivasis: They are the Scheduled Tribes in the Indian Constitution. Although they are not regarded as untouchables, they, along with the Scheduled Castes, are officially regarded as the most socially and economically disadvantaged groups in India.
6 Please see 'Enrolment by Social Category', Report ID 4004, UDISE+ Reports, Available at: https://dashboard.udiseplus.gov.in/#/reportDashboard/sReport

References

Acker, J. (1992). Gendering Organizational Theory. In A. J. Mills and P. Tancred (eds), *Gendering Organizational Theory*, Sage: Thousand Oaks, CA, 248–260.

Adams, Michele, and Coltrane, S. (2005). Boys and Men in Families: The Domestic Production of Gender, Power, and Privilege. In Michael S. Kimmel, Jeff Hearn and R.W. Connell (eds), *Handbook of Studies on Men & Masculinities*, Sage: New Delhi, 230–248.

AISHE Report2021-2022. (2022). *All India Survey on Higher Education*. Government Of India, Ministry of Education, Department of Higher Education: New Delhi. Available at: https://aishe.gov.in/aishe-final-report/ Last accessed on 30th August 2024.

Allendorf, K. and R.K. Pandian. (2016). The Decline of Arranged Marriage? Marital Change and Continuity in India. *Population and Development Review* 42(3): 435–464. https://doi.org/10.1111/j.1728-4457.2016.00149.x

Alvesson, M., Y. Billing. (2009). *Understanding Gender and Organizations*. New Delhi: Sage.

Bansal, P. (2013). *Youth in Contemporary India. Images of Identity and Social Change*. Springer: India.

Basham, A.L. (1967). *The Wonder That Was India*. Rupa and Co.: New Delhi.

Basu, A. (1991). The Indian Response to Scientific and Technical Education in Colonial Era 1820–1920. In *Science and empire: Essays in Indian Context, 1700–1947*, Anamika Prakashan: Delhi, 126–138.

Basu, A. (1999). Women's Education in India: Achievements and Challenges. In Bharati Ray and Aparna Basu (eds), *From Independence Towards Freedom: Indian Women Since 1947*, Oxford University Press: New Delhi, 135–157.

Basu, A. (2005). A Century and a Half's Journey: Women's Education in India, 1850s to 2000. In Bharti Ray (ed), *Women of India: Colonial and Post-colonial Periods*, PHISPC, Centre for Studies in Civilizations: New Delhi, 183–207.

Belliappa, J. (2013). *Gender, Class and Reflexive Modernity in India*. Basingstoke: Palgrave Macmillan.

Bem, S.L. (1993). *The Lenses of Gender: Transforming the Debate on Sexual Inequality*. Yale University Press: New Haven.

Bhog, D. (2002). Gender and Curriculum. *Economic & Political Weekly* 37(17): 1638–1642.

Burton, A. (2009). Contesting the Zenana: The Mission to Make Lady Doctors for India, 1874–1885. In Neelam Kumar (ed), *Women and Science in India: A Reader,* OUP: New Delhi, 21–54.

Chadha, G., and Achuthan, A. (2017). Feminist Science Studies. Intersectional Narratives of Persons in Gender-Marginal Locations in Science. *Economic & Political Weekly* 52(17): 33–36.

Chakravarti, U. (1993). Conceptualizing Brahminical Patriarchy in Early India. *Economic and Political Weekly* 27(14): 579–585.

Chakravarti, U. (2004). Conceptualizing Brahamanical Patriarchy. In M. Mohanty (ed), *Class, Caste and Gender*, Sage: New Delhi, 271–295.

Chanana, K. (1993). Accessing Higher Education: The Dilemma of Schooling Women, Minorities, Scheduled Castes and Scheduled Tribes in Contemporary India. In Philip G. Altbach and Suma Chitnis (eds), *Higher Education Reform in India*, Sage: New Delhi, 115–154.

Chanana, Karuna. (2001). *Interrogating Women's Education: Bounded Visions, Expanding Horizons*. Jaipur: Rawat Publications.

Chatterjee, Partha. (1989). Colonialism, Nationalism, and Colonialized Women: The Contest in India. *American Ethnologist* 16(4): 622–633.

Chiplunkar, G.M. (1930). *The Scientific Basis of Women's Education*. S.B. Hudlikar: Poona. Available at: https://archive.org/details/in.ernet.dli.2015.83360

Choksi, M. (1929). Some Impressions of Women's Colleges. In E.V. Gedge and M. Choksi (eds), *Women in Modern India*, D.B. Taraporevala Sons: Bombay, 63–77. Available at: https://archive.org/stream/in.ernet.dli.2015.200558/2015.200558.Women-In#page/n79/mode/2up/search/63. Last accessed on August 30, 2024.

Chowdhury, A. (2008). The Cultural Construct of Troubles with Boys in India. *Journal of Human Ecology* 24(4): 297–303.

Correll, S.J. (2001). Gender and the Career Choice Process: The Role of Biased Self-Assessments. *American Journal of Sociology* 106(6): 1691–1730.

Cronin, H. (2005). The Vital Statistics: Evolution not Not Sexism Puts Us at a Disadvantage in the Sciences. *The Guardian Weekly* 18-24 March 2005, 15.

Das, Upasak, and Singhal, Karan. (2023). Solving it correctly: Prevalence and Persistence of Gender Gap in Basic Mathematics in Rural India. *International Journal of Educational Development* 96(January 2023): 102703. Available at: https://doi.org/10.1016/j.ijedudev.2022.102703. Last accessed on August 30, 2024.

Datta, Anandita. (2005). MacDonaldization of Gender in Urban India. *Gender, Technology and Development* 9(1): 125–135. https://doi.org/10.1177/097185240500900107

Desai, S., and Andrist, L. (2010). Gender Scripts and Age at Marriage in India. *Demography* 47(3): 667–687.

Desai, N., and Krishnaraj, M. (2004). An Overview of the Status of Women in India. In Manoranjan Mohanty (ed), *Class, Caste and Gender*, Sage: New Delhi, 296–319.

Deshmukh-Ranadive, J. (2005). Gender, Power, and Empowerment: An Analysis of Household and Family Dynamics. In D. Narayan (ed), *Measuring Empowerment: Cross-Disciplinary Perspectives*, The World Bank: Washington, DC, 103–122.

Deshpande, Satish. (2003). *Contemporary India: A Sociological View*. Penguin Books: India.

Diwakar, Vaishali. (2016). 'It's Complicated': The Construction of Indian Middle-Class Teens in Social Media. *Journal of Creative Communications* 11(2): 161–182.

Donner, H. (2011). *Being Middle-Class in India: A Way of Life*. Routledge, Routledge Contemporary South Asia Series: London.

Dube, L. (2001). *Anthropological Explorations in Gender: Intersecting Fields*, New Delhi: Sage Publications.

Edström, Jerker, Das, A., and Dolan, C. (2014). Introduction: Undressing Patriarchy and Masculinities to Re-politicise Gender. *IDS Bulletin* 45(1): 1–10.

Eliot, L., Ahmed, A., Khan, H., Patel, J. (2021). Dump the "Dimorphism": Comprehensive Synthesis of Human Brain Studies Reveals Few Male-Female Differences Beyond Size, *Neuroscience & Biobehavioral Reviews*, 125 (June 2021): 667-697.

Etzkowitz, H., and Gupta, N. (2006). Women in Science: A Fair Shake? *Minerva* 44(2): 185–199.

Fernandes, L. (2006). *India's New Middle Class: Democratic Politics in An Era Of Economic Reform*. University of Minnesota Press: Minnesota.

Forbes, Geraldine. (2008). Education for Women. In Sumit Sarkar and Tanika Sarkar (eds), *Women and Social Reform in Modern India: A Reader*, Permanent Black: New Delhi, 83–112.

Forbes, G. (2009). No Science for Lady Doctors in Nineteenth Century Bengal. In Neelam Kumar (ed), *Women and Science in India: A Reader*, OUP: New Delhi, 3–20.

Fuller, C., and Narasimhan, H. (2007). Information Technology Professionals and the New-Rich Middle Class in Chennai (Madras). *Modern Asian Studies* 41(1): 121–150. https://doi.org/10.1017/S0026749X05002325

Gautam, Meenakshi. (2015). Gender, Subject Choice and Higher Education in India: Exploring 'Choices' and 'Constraints' of Women Students. *Contemporary Education Dialogue* 12(1): 31–58.

Gorringe, H. (2018). Afterword, Gendering Caste, Honor, Patriarchy and Violence. *South Asia Multidisciplinary Academic Journal* [online]. Available at: https://journals.openedition.org/samaj/4685. Last accessed on August 30, 2024.

Gupta, Namrata. (2007). Indian Women in Doctoral Education in Science and Engineering: A Study of Informal Milieu at the Reputed Indian Institutes of Technology. *Science Technology and Human Values* 32(5): 507–533.
Gupta, Namrata. (2016). Perceptions of the Work Environment: The Issue of Gender in Indian Scientific Research Institutes. *Indian Journal of Gender Studies* 23(3): 437–466.
Gupta, Namrata. (2017). Gender Inequality in The Work Environment: A Study of Private Research Organizations in India. *Equality, Diversity and Inclusion: An International Journal* 36(3): 255–276.
Gupta, Namrata. (2020). Patriarchy Reinforced or Challenged? A Study of Engineering Students in An Elite Indian Institute. *Gender, Technology and Development* 24(2): 250–270.
Gupta, Namrata, and Sharma, Arun K. (2003). Patrifocal Concerns in the Lives of Women in Academic Science: Continuity of Tradition and Emerging Challenges. *Indian Journal of Gender Studies* 10(2): 279–305.
Halpern, D.F., Benbow, C.P., Geary, D., Gur, R., Hyde, J.S., and Gernsbacher, M.A. (2007). The Science of Sex Differences in Science and Mathematics. *Psychological Science in the Public Interest* 8(1): 1–51.
Hearn, J. (2004). From Hegemonic Masculinity to the Hegemony of Men. *Feminist Theory* 5(1): 49–72.
Iyer, Padmini. (2014). *Negotiating Masculinities and Learning To 'Be A Man' At School in New Delhi, India.* A paper presented at the 2nd MenEngage Global Symposium: New Delhi, November 11, 2014. Available at: https://sussex.figshare.com/articles/presentation/Negotiating_masculinities_and_learning_to_be_a_man_at_school_in_New_Delhi_India/23419358?file=41255097. Last accessed on August 30, 2024.
JEE Report. (2018). *JEE Advanced Report 2018. Joint Implementation Committee Report. Volume I.* IIT: Kanpur. August 2018. Available at: https://www.iitk.ac.in/new/data/jee-report/1_JEE_2018_Vol_1.1_1.2-28-2-19.pdf. Last accessed on August 30, 2024.
Jodhka, S. (2015). *Caste in Contemporary India.* Routledge: India.
Kawale, V., and Maji, S. (2022). Exploring Science-Related Implicit Gender Stereotypes with Draw-a-Scientist-Test: A Study of Indian School Children. *Proceedings of the 2nd International Conference on Gender Studies and Sexuality*, July 29–31, 2022, The Republic of Ireland: Dublin. Available at: https://www.dpublication.com/abstract-of-2nd-icgss/51-19500/. Last accessed on August 30, 2024.
Kondaiah, B. K., Mahadev, S., and Wahlang, M.G.T. (2017). The Production of Science Bearing Gender, Caste and More. *Economic & Political Weekly* 52(17): 73–79.
Kumar, N. (2011). The Middle-Class Child: Ruminations on Failure. In A. Baviskar and R. Ray (eds), *Elite and Everyman: The Cultural Politics of the Indian Middle Classes,* Routledge: New Delhi, 220–245.
Kurup, Anitha, Maithreyi, R., Kantharaju, B., and Godbole, Rohini. (2010). *Trained Scientific Women Power: How Much Are We Losing and Why?* IAS–NIAS Research Report. Available at: https://cms.nias.res.in/sites/default/filesefs/2022-01/IAS-NIAS-Report%20%281%29.pdf. Last accessed on August 30, 2024.
Mandate Document. (2022). *Guidelines for the Development of National Curriculum Framework*, NCERT, April 2022. Available at: https://ncert.nic.in/pdf/Mandate-NCF.pdf. Last accessed on August 30, 2024.
Mazumdar, V. (2012). *Education, Equality and Development: Persistent Paradoxes in Indian Women's History.* Pearson: New Delhi.

Mazumdar, M. (2017). Science and the Making of a New Nationalist Masculinity. In S. Krishna and G. Chadha (eds), *Feminists and Science*, vol. 2, Sage: New Delhi, 175–206.
Mukhopadhyay, C.C. (1994). Family Structure and Indian Women's Participation In Science And Engineering. In C.C. Mukhopadhyay and S. Seymour (eds), *Women, Education and Family Structure in India*, Westview Press: Boulder, CO, 103–134.
Mukhopadhyay, C. C., and S. Seymour. 1994. 'Introduction and Theoretical Overview'. In C. C. Mukhopadhyay and S. Seymour (eds), *Women, Education and Family Structure in India*, Westview Press: Boulder, CO, 1–33.
NAS. (2018). *National Achievement Survey. National Report Card. Learning Achievements of Students, Class X (Cycle 2).* Department of School Education and Literacy, Ministry of Education, Government of India. NCERT 2021. Available at: https://nas.gov.in/report-card/2017
Nimkoff, M.F. (2005). The Family in India: Some Problems Concerning Research On The Changing Family in India. In T. Patel (ed), *The Family in India: Structure and Practice,* Sage: New Delhi, 71–80.
NSO. (2019). *Time Use in India 2019.* Time Use Survey, January–December 2019. National Statistical Office, MoSPI, Government of India. Available at: https://www.pib.gov.in/PressReleasePage.aspx?PRID=1660028. Last accessed on August 30, 2024.
Paliwal, A. (2023). How India's Caste System Limits Diversity in Science — In Six Charts. *Nature* 63(12 January): 230–234.
Powell, Abigail, Dainty, A., and Bagilhole, B. (2012). Gender Stereotypes Among Women Engineering and Technology Students in the UK: Lessons from Career Choice Narratives. *European Journal of Engineering Education* 37(6): 541–556.
PRICE (2022). *Gearing Up for a Billion-Plus Middle Class by 2047.* People Research on India's Consumer Economy. ICE360° Survey. Available at: https://www.ice360.in/app/uploads/2022/11/middle-class-brochure-2022.pdf. Last accessed on August 30, 2024.
Radhakrishnan, S. (2009). Professional Women, Good Families: Respectable Femininity and the Cultural Politics of A 'New' India. *Qualitative Sociology* 32(2): 195–212.
Radhakrishnan, S. (2011). *Appropriately Indian: Gender and Culture in a New Transnational Class.* Orient Blackswan: New Delhi.
Raina, S. (2012). Gender Bias In Education. *International Journal of Research Pedagogy and Technology in Education and Movement Sciences* 1(2). Available at: https://ijems.net/index.php/ijem/article/view/10. Last accessed on August 30, 2024.
Raj, K. (1991). Knowledge, Power and Modern Science: The Brahmins Strike Back. In Deepak Kumar (ed), *Science and Empire: Essays in Indian Context, 1700–1947*, Anamika Prakashan: Delhi, 115–125.
Rakshit, S., and Sahoo, S. (2020). *Biased Teachers and Gender Gap in Learning Outcomes: Evidence from India* (Working Paper No. 684). GLO Discussion Paper. Available at: GLO Discussion Paper website: https://www.econstor.eu/handle/10419/225065. Last accessed on August 30, 2024.
Ram, Usha, Strohschein, Lisa, Gaur, Kirti. (2014). Gender Socialization: Differences between Male and Female Youth in India and Associations with Mental Health, *International Journal of Population Research*, 357145. Available at: https://doi.org/10.1155/2014/357145. Last accessed on November 13, 2024.
Raveendran, A., and Chunawala, S. (2015). Reproducing Values: A Feminist Critique of a Higher Secondary Biology Textbook Chapter on Reproductive Health. *Indian Journal of Gender Studies* 22(2): 194–218.

Ridgeway, C.L., and Correll, S.J. (2004). Unpacking The Gender System: A Theoretical Perspective on Gender Beliefs and Social Relations. *Gender & Society* 18(4): 510–531.

Sancho, D. (2015). *Youth, Class and Education in Urban India: The Year that Can Break or Make You.* Taylor Francis Ltd: Milton Park.

Singh, J.P. (2010). India's Changing Family and State Intervention. *The Eastern Anthropologist* 63(1): 17–40.

Siqueira, T.N. (1939). *The Education of India: History and Problems.* OUP: Madras. Available at: https://archive.org/details/in.ernet.dli.2015.507462. Accessed on August 30, 2024.

Sondhi, G. (2015). Indian International Students: A Gender Perspective. In S.I. Rajan (ed), *India Migration Report 2015: Gender and Migration*, Routledge: New Delhi, 104–119.

Sridharan, E. (2004). The Growth and Sectoral Composition of India's Middle Class: Its Impact on the Politics of Economic Liberalization. *India Review* 3(4): 405–428.

Srivastava, A.K. (1997). The Changing Place of Child In Indian Families: A Cross-Generational Study. *Trends in Social Science Research* 4(1): 191–200.

S&T Indicators. (2023). *S&T Indicators Tables 2022–2023.* March 2023. Department of Science & Technology. Government of India: New Delhi. Available at: https://dst.gov.in/document/reports/st-indicators-tables-2022-23. Last accessed on August 30, 2024.

Subrahmanyan, L. (1998). *Women Scientists in the Third World: The Indian Experience.* Sage: New Delhi.

Subramanian, J. (2007). Perceiving and Producing Merit: Gender and Doing Science in India. *Indian Journal of Gender Studies* 14(2): 259–284.

Sur, Abha. (2009). Dispersed Radiance. Women Scientists in C.V. Raman's Laboratory. In Neelam Kumar (ed), *Women and Science in India*, OUP: New Delhi, 98–136.

Thapan, Meenakshi. (2007). Adolescence, Embodiment and Gender Identity: Elite Women in a Changing Society. In Rehana Ghadially (ed), *Urban Women in Contemporary India: A Reader*, Sage: New Delhi, 31–45.

Thapar, Romila. (1966). *A History of India 1.* Penguin Books: England.

UNESCO. (2021). Transforming Mentalities: Gender Equality and Masculinities in India. C. Coley, Christie M. Gressel and Rao R. Bhavani (eds). UNESCO: New Delhi. Available at: https://unesdoc.unesco.org/ark:/48223/pf0000377859.locale=en. Accessed on August 24, 2024.

Venkatesh, Sujatha. (2015). Forms of Social Asymmetry and Cultural Bias: Of Gender and Science in India and the World. *Transcience* 6(1): 1–19.

West, C., and Zimmerman, D. (1987). Doing Gender. *Gender and Society* 1(2): 125–151.

Women and IT Scorecard-India. (2018). *NASSCOM.* The Open University: UK. Available at: https://www.researchgate.net/publication/328603083_Women_and_IT_Scorecard_-_India_2018. Accessed on August 30, 2024.

1

GENDER AND HIGHER EDUCATION

Introduction

A series of educational choices from senior secondary school until a student enrolls in the undergraduate program requires careful planning and family investment. This chapter analyzes how the choices of students at this stage, and later those related to higher studies, are based on patriarchal considerations, creating gendered pressure on students. Such decisions mark the nascent gendering in the STEM fields.

Gendered subject choices: pre-college

Senior secondary education or Class XI and XII in school (for 16–18 year olds) constitutes a critical period in the career of a student. This stage occurs just prior to higher education or college education in India. A student in these two years (Class XI–XII) chooses a stream of learning which largely decides his/her subject options for college and, in turn, his/her career options for the future. There are three primary streams in most schools at senior secondary level: one, arts/social sciences; two, science (PCM, that is physics, chemistry, and mathematics; or PCB, that is physics, chemistry, and biology); and three, commerce. Here, our focus is on the science streams.

Students are required to study PCM and PCB to sit for engineering and medical entrance exams, respectively. Students might opt for biology along with PCM to exercise the option to sit for both exams. Thus, the students of Class XII, wanting to pursue a professional education, such as in medicine or engineering, not only take final class exams conducted by the national or state boards but also sit for separate entrance examinations (called

DOI: 10.4324/9781003596660-2

NEET-UG and JEE for medicine and engineering, respectively). Those unable to pass these entrance exams after Class XII can also sit for them while pursuing undergraduation in science.

The science stream offers more future subject options than any other stream. Thus, those in the science stream can shift to arts or commerce in undergraduation (UG), but the opposite is not possible; those in the commerce stream can move to arts but not to science. Science is considered the best option and humanities is the last option, while commerce is gradually becoming popular (Gautam 2015). The science stream, especially PCM, is the most popular and prestigious (NCEE 2005).

The enrollment figures of boys and girls in various science streams at the senior secondary level are not available. Figures for enrollment in undergraduation show that the proportion of women is higher than men in medical sciences (57.6%) and sciences (51%), while it is lower (29%) in engineering and technology (AISHE Report 2021–2022 Table 12). 'Sciences' might include those from PCM or PCB streams. These figures indicate that the proportion of girls opting for the PCB stream in senior secondary school might be higher than those opting for PCM. Women's education in science, and more remarkably in the engineering field (an offshoot of the PCM stream), has witnessed a growth in participation over the years. The proportion of women enrolled in engineering and technology courses at the undergraduate level was 0.2% in 1950–1951, which rose to 3.8% in 1980–1981, 16% in 1995, and 29% in 2020–2021 (Gupta 2012; AISHE Report 2021–2022: Table 12). However, the subject choices reflect both continuity and change in the Indian socio-cultural context.

In the Indian patriarchal system, family plays an important role in educational decisions (Mukhopadhyay 1994), such as those related to choosing a stream in school, choosing a college, and choosing a course of study in college; a father's role in such decisions is particularly significant (Gautam 2015). Studies (Gupta 2012; 2015; Varma 2011) show considerable family support for women that opt for the science stream in Class XI and for science/engineering later during higher education. Considerations of parents in making educational decisions are often gendered (Gupta 2012). For instance, the subject to be chosen is related not only to the family decision of whether the daughter should have a career-oriented education but also to the marital obligations of the parents. Thus, in accordance with the rules of hypergamy, parents must find a groom who is more educated and older than their daughter. In many castes and religions, parents of a professionally educated woman would find it difficult to find a groom who was more educated than their daughter and who would also 'allow' her to pursue a career. In the early 1990s, Mukhopadhyay (1994) found parents were unwilling to expend resources on their daughters' academic career due to the assumption that, after marriage, the benefits of their education would accrue to their husband and his family.

However, in recent years, a professional degree eases marital negotiations for a woman's parents and also empowers the daughters (Gupta 2015). A daughter's professional education brings prestige to the young woman and her family and assures her a respectable job. Change has been particularly visible since the late 1990s due to liberalization and globalization. An engineering degree has become an attractive option for girls due to its association with computer-related branches. The latter are considered 'appropriate' for girls (Varma 2010), and access to engineering education has increased due to privatization of education and a mushrooming of engineering colleges (Gupta 2012). About 75% of all engineering- and technology-approved seats are in private institutes, including self-financing, private deemed, and state private universities (AICTE Statistics 2024). Further, since liberalization, the availability of jobs in this field has risen (Gupta 2015). A PCM stream is also desirable for those wishing to take up science courses later. Gupta's (2012) study of a local women's engineering college found that parents did not wish their daughters to study humanities or social sciences in Class XI–XII. The parents do not necessarily provide any JEE coaching to their daughters, which is expensive, but support their tuitions for the PCM subjects. However, the stream requires the study of mathematics. Considering that math is thought to be a masculine domain in the West (e.g., Brandell and Straberg 2008), the growth of women's participation in engineering has ignited debates among scholars as to whether women's mathematical abilities have ever been doubted in India.

Contrary to the Western experience, the beliefs about women's incompetency in mathematical or mechanical activities or notions of differences in intrinsic intellectual ability do not seem to appear in the Indian socio-cultural context (Mukhopadhyay 2009). Varma (2011), in the context of computer education for girls in India, also found that mathematics does not appear as a male domain and that gender does not play any role in acquiring mathematical and problem-solving skills. A study on women in a low-ranked state engineering college for women (Gupta 2012) found that fathers of 96% of respondents and mothers of 93% of respondents believed that women were no less capable than men in science and math. At present, the proportion of women enrolled in Master degrees in Mathematics exceeds that of men at about 62%, and in PhDs it is about 50% (AISHE Reports 2021–2022).

However, there is evidence to the contrary also, and certain studies indicate that doubts regarding women's ability to do well in mathematics continue to endure. For instance, Gautam (2015) found that boys and teachers questioned the mathematical ability of girls in senior secondary school. This was supplemented by gendered classroom experiences, as boys received greater attention in the 'masculine' subjects of mathematics and physics than girls did; stereotypes that girls are good in rote learning and boys in conceptual understanding tend to persist.

Evidence as to a gender gap in math ability shows considerable geographical variations, indicating socio-cultural roots of the gap. A national survey found that average mathematics scores for girls and boys of Class X do not show any statistically significant difference (NAS 2018). However, Das and Singhal (2017) found a gender gap in mathematics at the primary school level in rural areas and not in urban areas. The rural gender gap in mathematics scores was a product of societal factors, such as parents discussing science and related issues more with sons than with daughters, involvement of girls with household chores, boys' involvement in outdoor activities, and lack of nutrition among girls. The study does not report any intrinsic or biological differences being perceived by society. Another study by these authors (Das and Singhal 2023) found significant interstate variation, with most of the Northern states exhibiting disproportionately larger gender gap in mathematics performance, while the Southern Indian states were showing a 'reverse gender gap' (where boys are performing worse than girls) in rural areas. Thus, the evidence clearly negates not only any biological explanation but also that belief in innate differences is universal.

Within the sciences, there is a greater interest among women in biological sciences than in physical sciences. The interest in biological sciences among women is a function of socio-cultural experiences. For instance, a study (Gafoor 2017) shows that despite their similar achievement levels, boys at the senior secondary level had a higher interest in physics, and girls had a higher interest in biology and chemistry. These interest levels, according to the author, related more to their out-of-school experiences rather than innate abilities.

The innate ability or the academic/career interests of a son/daughter appear secondary to the primary role of gendered considerations in the family's subject choice decisions. These decisions are gendered for boys also. According to Gautam (2015: 37), in urban middle-class families, 'even if brothers were not considered "good students", they still "had to take science". Male siblings were not given the option of pursuing arts and commerce'. Further, in such families, the decision to take science was made by keeping in mind mainly two professions: medicine for girls and engineering for boys. Although women's participation in engineering has increased, this is mainly owing to the computer-related fields of engineering. Engineering is still male-dominated with only 29% of women enrolled in undergraduation. A higher proportion of men in engineering is linked with the social construction of masculinity and male privilege that celebrates men's academic achievements, the latter being significant for families to settle the son as a successful breadwinner.

The entry in the two sought-after professions of engineering and medicine requires extensive preparation which has led to coaching centers. The trends in coaching reflect family choices that vary for sons and daughters. This gendered nature of coaching and a medical/engineering decision is discussed in the following section.

'Coaching' for entry in medicine and engineering: gendered decisions

Since 1991, there has been a mushrooming of private coaching centers. Education is the fulcrum of middle-class aspirations, and coaching centers for the final school year (Class XII) symbolize the mobility aspirations of this class (Sancho 2015). This is because the Class XII examinations constitute the bedrock of future higher education and jobs. The entrance exams for medicine/engineering require coaching at the senior secondary school level because they demand training and skills of a higher level than those required for the Class XII examinations. Coaching is not essential, but it enhances the chances of securing admission. For instance, the JEE-Advanced Report (2018) shows that out of 60,420 of those who registered for the JEE-Advanced exam (for seats in IITs) and took coaching, about 10.5% (6,326) finally received admission into an IIT; while out of 97,253 students who registered and did self-study, only 5.32% (5,183) finally received admission (Ch. 14/1).

Studies show a lack of female students at the coaching centers for engineering entrance examinations. For instance, Singh and Pathak (2010) found that more women than men were not being given professional coaching for such examinations. Further, almost all the students who did well in JEE had taken coaching, but the percentage of women in coaching centers in major cities of India did not exceed 24%. Gupta's (2012) study corroborates this, as out of about 58% of the respondents at a local private women's engineering college who sat for the JEE (equivalent to JEE-Advanced at the time of this study), only 30% had taken coaching.

Few women sit and qualify in the engineering entrance exams. The JEE Mains, conducted twice nationwide annually, provides scores for entrance to engineering colleges other than the IITs. Women constituted one-third of those who sat for the JEE Mains in 2023.[1] Of 1.89 lakhs candidates that registered for the JEE (Advanced) 2023 examination, about 23% were women; and of those who qualified, 17.2% were women (JEE-Advanced Report 2023). This low percentage of girls that qualify partly suggests a lack of coaching since girls are well represented in the merit lists of Class-XII CBSE boards and are often toppers (obtaining highest marks) in the PCM stream (Singh and Pathak 2010; Anand 2023). It also suggests a lack of expectations from girls, and such families that do not impart coaching usually prefer to send their daughters to a nearby engineering college (Gupta 2012). Keeping these factors in mind, and an extremely low proportion of women in the IITs in 2016 (8%), a 20% supernumerary quota for women was introduced in the IITs from 2018, which was fully achieved in 2021–2022.[2]

The situation in the medical entrance exam is radically different, with girls' participation exceeding that of boys'. The selection of students in most of the medical colleges is based on the score obtained in the objective type exams, such as NEET-UG for admission into undergraduate medical courses at government and private medical colleges. Coaching is believed to be desirable for obtaining a good rank in the merit list, and thereby a good medical college; the exam is regarded as tough and enjoys prestige similar to the engineering entrance exam (JEE). There is a lack of studies on gender issues related to medical coaching. However, unlike the JEE exams, girls' proportion in medical entrance exams is higher than that of boys'. For instance, in 2023, of those who sat for the NEET-UG, about 56.7% were female and females constituted the same proportion among those who qualified.[3]

According to the head of a coaching institute at Hyderabad:

> For medical coaching, we see six girls to three boys. In engineering, it is the reverse. Engineering has a lot of mathematics in the curriculum. The general notion is that girls are more suited for studying biology.
>
> *(Deccan Chronicle 2017)*

This leads to the questioning of the hypothesis that parents are unwilling to spend money on the coaching of girls. The monetary investment for girls' education depends on the subject choice decision. Engineering continues to be viewed less favorably than medicine for girls. The preference for a medical career for girls is due to various gendered factors. According to Chakrabarty (2010: 420):

> Parents usually prefer the professional career of a doctor (compared to engineering) for their daughters, as it offers flexible working hours, independent work through private practice, social prestige, and good earning.

As apparent from the proportions of those sitting for the national entrance exams, a lower proportion of men (less than half) than women sat for the medical entrance (NEET-UG), and the situation is reversed for the JEE Mains (two-thirds of men). The question then arises: why is engineering preferred for men and not medicine? A successful medical career requires a greater period of study including five-and-a-half years of bachelor's (MBBS), three years of master's and maybe two years of super-specialization; while an MBBS (Bachelor of Medicine, Bachelor of Surgery) degree holder can technically practice medicine, the degree has little market value unless supplemented by the master's degree. On the other hand, four years of a Bachelor in Engineering can fetch a job with a decent salary, which is further enhanced by a two-year Master degree in Management (Rashmi 2005). Obtaining a good rank in the JEE and NEET-UG are equally difficult, but entry in IITs or other reputed public and private engineering institutes guarantees a good career, while a medical graduate would still have to slog it out for several more years. A doctor needs to establish a reputation, but once established, can make a higher amount of money than an engineer. A successful doctor lacks time for vacation and family. It is easier to go abroad for higher studies after a Bachelor in Engineering than after an MBBS.

In short, it implies that a successful career is easier to achieve through engineering than medicine. Yet, a career in medicine is preferred for girls. A study of medical college students shows that women received more encouragement from their families to pursue medicine than men (Agrawal et al. 2015). It appears contradictory that women receive encouragement for medicine from their families even though it is difficult to achieve a successful career in medicine. However, preference for medicine for women is due to historical and socio-cultural reasons. Historically, medicine has been considered prestigious for women since the colonial period, and being related to biology it appears to be suited to women's aptitude. Further, encouraging daughters for a career in medicine is also probably because of various other gendered factors: highly successful careers are not desired for women, hence, the difficulties in attaining a successful medical career concern families with sons more than those of daughters. Studies suggest higher parental expectations from boys for academic and career achievements (Basu et al. 2017; Larsen 2011). Women are expected to use their education to be independent financially, while men are expected to be 'successful' in careers (Gupta 2007). As Kapadia (2017: 153) puts it, girls in middle-class, urban families are raised to be achievement-oriented and their aspirations encouraged, but they are persuaded to exercise restraint so that their aspirations do not result in a compromise of their gendered roles in family and home. The ideas of educating and empowering women versus placing traditional limitations on those coexist (Radhakrishnan 2011). In effect, it implies that a woman might become a doctor as it enables financial independence and status, but the practice of medicine or the choice of specialization should allow her to meet the gendered obligations. A private practice at home and flexible work hours, as suggested by Chakrabarty (2010), would help in balancing home and work.

The coaching at senior secondary level has become a menace, a 'shadow education' (Orberg 2017). Coaching represents a failure of the institutions to select the best students for professional courses. Kota in Rajasthan is the largest coaching market in the country. The Kota coaching industry has about 110,000-strong student population with a yearly turnover, including fees and living costs, around €200 million (Orberg 2017). They have a large number of successful qualifiers in the JEE and NEET. It also has the highest suicide rates among the students with about 15 students committing suicide every year (Poonam 2016; Abraham 2022); and the number worryingly surpassed that in 2023, with the toll going up to 23 at the end of August (Sengupta 2023). Student suicides have also been reported from other coaching centers such as those at Andhra and Telengana (Dailybite 2017). The premium attached to being doctors and engineers and a lack of other respectable jobs fuel parental pressure. The cost of coaching borne by parents creates immense performance pressure on students, usually from the middle class. A grueling study schedule, a culture of glorifying toppers, a large student population in the coaching class, difficulty in coping with the preparation, lack of friends, and cultural differences with peers might lead to isolation and depression.

The stress in the two years of higher secondary school affects both boys and girls (usually 16–19 years old). National Crime Records Bureau data (NCRB 2021) found that of total suicides among students and unemployed (13,089), 43.5% are female. Perhaps unemployment is a greater worry for men than for women in the Indian context, which accounts for a higher proportion of suicides among male students and unemployed. In a study on academic stress among students of Class XI–XII (Deb et al. 2015), about 63% of students reported stress due to academic pressure with no significant differences across gender. About 66% of students reported pressure due to parental expectations for better academic performance. Although there is a higher parental pressure on boys than girls, the difference is not significant (Deb et al. 2015: Table 2). Perhaps both sexes experience academic pressure due to parental expectations, for boys to obtain a lucrative career and for girls to obtain high status jobs, to meet patrifocal concerns of parents (e.g., see Gupta 2012).

The latest spate of suicides has prompted debates about the measures to be taken by the coaching centers (Sengupta 2023). The suicides are partly responsible for a substantial decline in the student population in Kota centers in 2024 (Bansal 2024). However, coaching centers continue to flourish elsewhere, and the major coaching centers of Kota have expanded to other cities. The combination of parental expectations, state of education (few good institutes), and economy (few lucrative jobs) are responsible for the mad craze for engineering/medicine, and hence, for coaching. The long-term solutions require creating greater awareness of the diverse jobs available, parent and student counseling, and establishing a greater number of high-quality educational institutes (for engineering/medicine).

Selection after class XII: masculine/feminine branches

The enrollment proportions in undergraduate courses in STEM indicate horizontal segregation. Of the total number of students enrolled in the undergraduate (or UG) arts, social sciences courses, as well as science courses, about 50% are women; in medical sciences it is 57.6% (less in MBBS at about 49.3%); and in engineering and technology, 29% (AISHE Reports 2021–2022: Table 11, 12). The education stream is female-dominated, with about 63% women (Table 11: pt 37); Law is male-dominated, with about 33% women in the stream (Table 11, pt. 131).

Within science, no separate statistics are available for undergraduate enrollments in physics, chemistry, math, and bio-sciences. The 'leaky pipeline' in science education, which describes the phenomenon of more women than men dropping out of higher education as they move from undergraduation to PhD (Miller and Wai 2015), does not seem to apply to India. As Table 1.1 shows, there is an increase in the enrollment proportion of women from UG to master's and then drops again in doctoral studies for science. Engineering/technology exhibits a different trend in 2021–2022 compared to other years; there is an increase from UG to master's to PhD in women's enrollment proportion. The medical sciences' enrollment of women increased from UG to master's in 2021–2022, a pattern first observed in in 2020–2021.

In the last decade, there has been a jump in the proportion of women enrolled in sciences at all levels, UG, Master's, and PhD (Table 1.1). The cause of a high enrollment has been attributed to 'feminization' of sciences, as the globalization and market forces since the 1990s have made pure sciences less lucrative commercially, drawing fewer job opportunities compared to the other branches of science and engineering (Krishna 2001; Chanana 2001). Further, pure sciences are more prestigious compared to humanities, as those opting for science are perceived as bright students by society (Gautam 2015), enhancing the marriage prospects of daughters opting for it (Gupta 2007).

However, the proportion of women in the top science institutes that integrate education and research, and admit through nation-wide exams, is much less than the all-India figures. For instance, in 2022, at IISER-Pune, women constituted only 27.6% of the total in the UG Program (a five-year program of integrated BS-MS), while the prestigious IISc had only about 14.1% of women in its four-year UG (BS-Research), and 17.4% in its two-year master's program (NIRF 2024). This again suggests the relationship between gender and reputed institutes, entry into which requires coaching/ tuitions.

Further, feminization cannot explain the growing numbers at the doctorate level. For instance, the proportion enrolled in a Doctorate in Physics has increased by about 10% since 2010–2011. This substantial increase in sciences appears to be a product of government efforts, establishment of quality science educational institutes at the national level, and growing awareness among women about research careers.

Overall, in contrast to science, engineering is male dominated, with women constituting about 29% of enrollment in undergraduation, although the percentage is high compared to the Western countries. In terms of degrees

TABLE 1.1 Percentage of women to total enrollment in science and engineering at various levels since 2010–2011

		Percentage of women		
	Faculties	*Undergraduation (UG)*	*Master's*	*PhD*
2010–2011	Physics	–	46.45	29.54
	Chemistry	–	44.3	36.97
	Math	–	51.72	38.83
	Bio-Science	–	61.48	53.1
	All Sciences	46.51	52.01	39.84
	Engineering/Technology	29.15	31.83	23.46
	Medical Sciences (total)	53.5	42.68	36.50
2017–2018	Physics	–	57.8	35.8
	Chemistry	–	57.3	40.8
	Math	–	59.6	44.9
	Bio-Science	–	69.9	48.4
	All Sciences	48.6	61.2	45.6
	Engineering/Technology	28.6	37.4	30.8
	Medical Sciences (total)	61.1	57.1	43.3
2021–2022	Physics		54.5	40.2
	Chemistry	50.7	54.1	45.9
	Math		61.6	51.3
	Bio-Science		69.0	64.2
	All Sciences		61.2	50.0
	Engineering/Technology	29.1	32.2	34.0
	Medical Sciences (total)^	57.6	59.1	52.2
	Medicine*	49.3 (MBBS)	53.4 (MD)	31.5 (DM)

Source: AISHE Reports 2021–2022 ; 2017–2018; 2010–2011: Table 12, 13; *Table 11; (^includes all forms of medicine such as Ayurveda, homeopathy, Unani)

earned in engineering and technology, the proportion of women is about one-third in bachelor's (29.2%), postgraduate (35.5%), and PhD (29.1%) degrees (AISHE Reports 2021–2022, table 36). In the US, the proportion of women earning engineering degrees has been increasing but is a quarter of degrees at all levels (NSF 2023). A high proportion of females in engineering in India is due mostly to the popularity of the IT/computer sector and the conscious 'choice' for computer-related branches among the urban middle classes. Most private colleges offer only computer-related courses, such as information technology (IT), electrical and electronics engineering (EEE), electronics and communication engineering (ECE), and computer science and engineering (CSE) (Gupta 2012).

In engineering and technology, the women's proportion in doctoral studies has been rising. In fact, the decadal jump in doctoral enrollment is almost 10%. The proportion of women in Master's courses and doctoral programs of medical sciences has also increased, with the decadal jump in doctoral enrollment being about 16%. This, perhaps again, reflects a growing interest of women in STEM education and careers.

In general, there is evidence of a waning interest in engineering in recent years. Total enrollment in UG engineering has declined from about 42 lakhs in 2014–2015 to about 39 lakhs in 2021–2022 (AISHE Reports, Table 12). About 550 engineering colleges were shut down between 2015–2021.[4] There is a moratorium on establishing new engineering colleges until 2024 with some exceptions.[5] A lack of employment or suitable job, particularly in core branches (such as civil, mechanical, chemical) even after investing in coaching, low-skill development in local and private engineering colleges, and better opportunities with other undergraduate courses are likely to be some of the reasons. The decline in UG enrollment in engineering for women is lower at 4.5% between 2014–2015 to 2021–2022, compared to men (9% decline).

In contrast to the declining trend in engineering in general, computer engineering continues to be attractive, particularly with the growing demand for Artificial Intelligence, the Internet of Things, Big Data, Blockchain technology, etc. Thus, the total enrollment in undergraduate computer engineering has risen from about 7.4 lakhs in 2014–2015 to 12.5 lakhs in 2021–2022, an increase of 69% (AISHE Reports 2014-2015 and 2021-2022 Table 12). However, the rise of female enrollment in this field is only 25% (from 3.42 lakhs to 4.28 lakhs) compared to 105% among men (from 4.02 lakhs to 8.25 lakhs). Thus, whenever a field of learning or an educational institute is higher in the hierarchy of prestige, a lower proportion of women may be found.

In contrast to engineering, there appears to be a rising interest in medical undergraduate courses. Perhaps, 'massification' of engineering (Mathew 2022) due to the structural expansion of education and low employability of graduates (NER 2019) is degrading its merit and reinforcing the merit of studying MBBS. Although lower in numbers compared to engineering, the number of students enrolled for undergraduate medical sciences has risen continuously from about 8 lakhs in 2014–2015 to 17 lakhs in 2021–2022. The increase in enrollment in MBBS has almost doubled among women (94% increase) and by 89% among men in this period (AISHE Reports 2014-2015 and 2021-2022 : Tables 11, 32). The government plans to set up 100 new medical colleges by 2027 by upgrading district hospitals.[6]

PCB students can opt for various courses related to biological sciences apart from MBBS, such as dentistry, paramedical, veterinary science, marine biology, etc. However, on clearing the NEET-UG exam, one can opt for either MBBS or Bachelor of Dental Surgery (BDS). The top rankers usually prefer MBBS. Women constitute half the proportion in MBBS but are about 73% of the total in the less coveted dentistry (BDS) (AISHE Reports 2021–2022: Table 11).

Choice of institutes for engineering/medicine

Online centralized counseling decides the medical college one studies in, and the top rankers usually opt for the prestigious government colleges. Overall, women constitute a lower proportion in the UG (undergraduate) and PG (postgraduate) medical programs in the elite government institutes (Table 1.2) and higher in the elite private institutes, such as Christian Medical College (CMC) at Vellore.

There are elite women's medical colleges as well; for instance, Lady Hardinge Medical College is a prestigious women's college in Delhi. On the other extreme, the army medical colleges have a fixed seat ratio based on sex with a higher number of seats for men than women; for example, AFMC Pune had 115 seats for men and 30 for women in 2023 (AFMC Information Brochure 2023).

Having more than a 25% proportion of women in even the most prestigious government medical colleges (Table 1.2) is a striking contrast to the proportion of women in the elite engineering institutes that had to resort to a quota system to create a 20% proportion of women. Prior to the quota system, in 2016, women in IITs constituted only 8% of those offered admission (JEE-Advanced Report 2016: 11). Perhaps, the high proportion of women in medical education is due to higher proportions of women opting for, and taking coaching for medicine than engineering.

A higher preference for the institute than the course of one's choice involving the educational decisions of women is suggested in Gautam (2015), who found that at the undergraduate level, women's subject choice was compromised. Participants in her study gave priority to the location of the institute (nearby college preferred), to the availability of an attached hostel, and in some cases, to single-sex institute. Gupta's (2012) study of a local women's engineering college found that among the factors that influenced respondents into opting for this college, the most important aspect was that it was their parents' choice; and second, its proximity to their home. In the IITs, the supernumerary seats provide women a better chance of joining the IITs in close proximity to their homes and settling for their preferred course, according to a former director at IIT Delhi (Smart 2023).

TABLE 1.2 Proportion of women in elite medical institutes in 2023

	UG (5 yr)	*PG (3 yr)*	*NIRF Rank*
AIIMS (autonomous)	25.8%	38.9%	1
BHU (public)	26.2%	35.1%	7
CMC, Vellore (private)	53.6%	59.7%	3

Source: NIRF 2024

The proportion of women in reputed government institutes, next to the IITs, such as the National Institutes of Technology (NITs), where the supernumerary quota system for women does not apply, the proportion of women could be less than 20%.[7] Thus, in the NITs, such as NIT Rourkela, and Prayagraj, according to NIRF 2024, the proportion of women in undergraduate programs hovered around 18–19%. In the private institutes of repute, such as BITS-Pilani which conducts its own entrance exams, the proportion is only 13%; in a reputed provincial government college such as Jadavpur, it is only 10.4%. However, the proportion of women is above 20% in many private engineering colleges (NIRF 2024).

Private engineering/medical colleges have provided access to students who aspire for higher education in these fields but are unable to make it to the elite institutes, which are mostly government funded. The criteria for admission to most private professional institutes/colleges are less tough than the elite public institutes. While medical colleges for women existed even during the colonial period, several women's engineering colleges have also emerged since the 1990s. Cummins College of Engineering for Women at Pune, established in 1991, was the first such college in the country. At present there are about 151 women's engineering colleges in the country.[8]

The quality of education is questionable in many private and public engineering and medical colleges. Only a small percentage of private colleges show up in the ranking of the top 50 engineering colleges. More than 90% of the private engineering colleges are affiliated colleges that have little academic autonomy (Banerjee and Muley 2008). Similar is the case with private medical schools, which are 53.8% (205/381) of all the medical schools in 2013 (Solanki and Kashyap 2014), and their contribution towards any pedagogical innovation or excellence is negligible. However, quality is also suspect in many government-run colleges. Various reports, which usually include all types of colleges, point out the failure of the engineering education in general due to a huge employability gap (e.g., NASSCOM 2022; NER 2019).

Despite differences in the quality of education imparted in various colleges, there are certain factors common to all the students who enter engineering or medical undergraduate programs, which affirm considerable family support for such programs. For instance, studying science at Class XI-XII level demands dedication of considerable time and energy from a student. It requires considerable support from the family. The daughter needs to be freed of all household responsibilities. Resources are required to provide for tuition or coaching, without which competing in entrance exams of engineering/medical or even sitting for Class XII examinations (which are conducted by national or state boards), is tough. Finally, an undergraduate engineering or medical education is quite expensive in private professional engineering and medical colleges, and the fees are higher for private medical courses than engineering. For instance, MAHE (private institute) charges, for General Category students, a maximum of about INR 19.6 lakhs for BTech (Bachelor of Technology), a four-year course, but about INR 71 lakhs for MBBS, a five-and-a-half-year course.[9] It also has more than 50% of women enrolled in the MBBS program (NIRF 2024). This suggests that the course fee or expenditure on a professional education does not deter families from making a choice for their daughters to become doctors or engineers.

Postgraduation studies: science, medicine, and engineering

Women's participation in postgraduation courses in STEM is higher than many countries, and the leaky pipeline is not visible for many fields. However, there is horizontal segregation in STEM fields and sub-fields, and the choice of Masters' degree in engineering might be gendered.

The figures for women's proportion in PG and PhD in Physics, Chemistry and Mathematics compare well with the Western countries. For instance, in India, in 2016–2017, the proportion of women earning degrees in physics was about 39% and 59% in PhD and master's respectively; and about 50% of doctoral degrees in mathematics (AISHE Reports 2016–2017, Table 36). In the US, the proportion of women earning degrees in physics at all levels was about 20%, in mathematics it was about 30% at the doctoral level in 2014 (NSF 2017: 6).

Women continue to dominate bio-sciences at PG and PhD levels (Table 1.1). The proportion of women in medical sciences has arisen between 2010–2011 and 2021–2022 (Table 1.1) at all the levels from UG to PhD. Unlike some of the Western countries, the profession of medicine is not considered a male profession in India (Chanana 1988). The figures show that the leaky pipeline in medicine from UG to PG existing earlier has now halted. In fact, there is a slight increase in the proportion of women doing PG. This shows the growing importance of medical careers among women. However, the proportion of women drops from PG to doctoral studies (Table 1.1). Further, there is a horizontal segregation within medical fields, with a low proportion of women in general medicine, general surgery, orthopedics, gastroenterology, radiology, cardiology, and neurology, among others. There is no female in a doctorate program in orthopedics (AISHE Report 2021–2022). Women dominate gynecology, ophthalmology, nursing, dentistry, physiotherapy, and pathology at the PG level; and in almost all fields there are more men than women at the doctoral level. Studies show that male-dominated disciplines like surgery and orthopedics are more lucrative (INSA 2004). The preference of women for branches such as dentistry is worldwide and stems from the opportunity they provide for autonomy and flexibility to manage work hours and thus balance home and family (Pallavi and Rajkumar 2011). Certain branches, such as pathology, are paraclinical branches in which women have dominated for several decades (Nagarajan 2016). However, figures also indicate that more men than women are entering pediatrics in PG and PhD, though the field has been traditionally preferred by women (Sood and Chadda 2010).

Women's proportion in medicine is higher at the UG level than at the PhD level; on the other hand, in engineering, their participation is higher at the PhD level compared to the UG level. The proportion of doctorates (DM) earned by women in medicine (at 31.5%) falls substantially from the MD level (53.4%) also. This again shows that a doctorate in medicine requires a greater investment of time than in engineering (starting from the UG degree), and might be a difficult 'choice' for women.

Defying the leaky pipeline, there is an increase in the proportion of women from BTech to MTech (Table 1.1). A study (Thakkar et al. 2018) found that a Master (MTech) in Computer Science and Engineering for daughters was encouraged by the families, as it was a matter of honor and a boost to the daughters' marital profile. Degrees were at times printed on the wedding invitations and helped to negotiate dowry. Further, the MTech degree offered women more freedom in a socially acceptable way to postpone marriage or work. The increase in the proportion of women in having a master's degree could also be perhaps because teaching might be preferred for women, which is also indicated by Parikh and Sukhatme (2004). These authors found that about one-third of women with BTech were employed in technical educational institutes. Since then, there have been changes in the AICTE Norms; and as per 2010 AICTE regulations, a master's is essential to teach at entry level in technical institutions (AICTE Handbook 2012-2013: 120, pt 8 .2). Further, more men than women tend to get jobs after BTech due to gender segregation in the job market (Gupta 2015), and men are expected to start earning as soon as they graduate and are under pressure to build successful careers (Mittal 2013).

Also, the proportion of women in PhD enrollments in engineering has been growing over the years (Table 1.1). This seems to be part of the positive development of an increasing acceptance of women's higher education and careers in STEM (further discussed in Chapter 4).

Women's migration abroad for higher education

Indian students formed about 8% of mobile students throughout the world among the students going abroad for higher studies (second only to China), and the highest proportion (25%) of Indian students going abroad chose the US, followed by Canada and Australia in 2020 (UNESCO-UIS 2023). India accounts for the second largest outflow of students after China (Sondhi 2015). Postgraduation courses abroad are more sought after than undergraduation. Thus, of those Indian students going to the US in 2012, only 18% went for UG studies while 81% sought posgraduation in 2014 (EY-FICCI 2014: 15). Indians form the largest cohort of international students in the US in 2023-2024, with a 23% increase since the previous year according to the latest Open Doors 2024 report, and a larger growth in postgraduate students than undergraduates (Open Doors 2024). According to a study by Sondhi (2015), studying in better universities, having better job prospects, and mitigating the fear of failure in a highly competitive Indian education system were common reasons for studying abroad. Also, expanding incomes, economic growth, and rising tertiary enrollments are important reasons for growth in mobile postgraduates (British Council 2014).

Globally, women and men are equally represented in the mobile student community (Sondhi 2015). In China, women comprise 55% of its mobile student flow; in India, women comprise only 27% of the international student mobility flow (Sondhi 2015). There is a paucity of studies on the gender aspect of student migration. In a survey of students who had studied abroad, it was found that enrollment in STEM programs exhibited maximum male bias, with 69% male and 31% female enrollments (Sondhi 2015). In 2022, there was a 100–150% increase in the number of women applicants (in all fields of study) to study abroad compared to the pre-pandemic year of 2019; the male-female student ratio among applicants seeking admission abroad improved to 50:50 in metro cities in 2022, from 70:30 in 2019, while other towns are beginning to catch up with this ratio (Bhattacharyya 2023).

There are class differences and gendered concerns related to student migrations. Sondhi (2015) found that women who study abroad belong to a higher social class with highly educated parents working in high-income positions, while men belong to families from various socio-economic groups. Further, concerns about gender inequality in the Indian labor market, particularly in engineering and IT, motivated some women to opt for higher studies abroad; also, while parents sought to send their daughters abroad for a better future (with respect to work and marriage), parents of sons were reluctant due to concerns that their sons might not return (making future of elderly care uncertain). According to the author, this suggests a noticeable change in the gender role expectations for women; while men still have to grapple with traditional roles, a daughter's emigration for education is aimed at negotiating better marital prospects and an unequal labor market. Hercog and Laar (2017) found that parents' support (financial and moral) is crucial for moving abroad, and obligations towards family are considered in the first place in the children's decisions to move abroad. Thus, given the fact that daughters traditionally have fewer obligations towards their parental families than sons, Sondhi's findings indicate how Indian patriarchy can create difficulties for sons (in this case, in their decision to move abroad). In view of this, the greater male migration for STEM courses suggests masculinity of the field.

Conclusions

This chapter highlights how women's choice of science stream in higher secondary school, 'coaching', and later of educational institute, are a product of gendered discourse in the families. Urban, middle-class families are extensively involved in planning their child's professional future in terms of financial and emotional investment, which is evident in the preparation (or coaching) for the entrance exams of engineering and medical undergraduate programs. Son preference is accompanied by parental attention on sons for professional education. The process also constructs women as secondary to men in the discourse on successful careers.

The socio-cultural norms and those related to masculinity/femininity tend to be reproduced and reconstructed through choices/decisions regarding subjects and institutes. The male privilege is reproduced in a horizontal segregation within the STEM fields in education. Sciences are preferred for women compared to humanities; medicine is preferred over engineering due to patriarchal reasons. Within medicine there are 'masculine' (e.g., surgery, orthopedics) and 'feminine' fields (e.g., gynecology). Within engineering, women's enrollment in computer science and engineering is encouraged as it is considered 'appropriate' for women, while civil/mechanical engineering is considered masculine. In general, numbers of women are inversely proportional to the prestige of a course and institute. While the popularity of computer science and engineering is growing, the rise of female enrollment in undergraduation in this course is a mere 25% compared to 105% among men between 2015 and 2021. Movement from bachelor's to master's is marked by a higher proportion of women owing partly to the significance of the degree in the patrifocal scheme as witnessed in the case of the MTech degree.

The proportion of women in science courses is much less in the national institutes of repute, compared to the national figures. The higher the prestige of the institute, the tougher the admission process and the more the requirement for 'coaching'. However, due to the gendered perception of the fields, 'coaching' daughters for medical education appears to be less of a constraint compared to that for an engineering education. Further, families might prefer local colleges for daughters due to the proximity to home, thus foregoing the opportunity of a more prestigious college.

There have been some positive trends in recent years. Affirmative action has brought women's numbers to one-fifth at the UG level in the prestigious IITs, the impact of which is yet to form a subject of sociological research. Proportions of women at all degree levels in science, technology, engineering, and medicine have risen since 2010–2011. However, women's rising proportions in STEM education fail to translate to careers in the same proportions, as discussed in the later chapters.

Apart from families, there are systemic issues that feed into the societal stereotypes and construct fields as 'masculine' or 'feminine'. This is analyzed in the next chapter with reference to the engineering education.

Notes

1 National Testing Agency: https://nta.ac.in/Download/Notice/Notice_20230502163550.pdf
2 See: https://www.pib.gov.in/PressReleasePage.aspx?PRID=1883470.
3 National Testing Agency: https://static-ai.asianetnews.com/common/01h2tz40kt30fdqh9yb72b6m1w/press-release-for-final-nta-score-for-neet--ug----2023.pdf
4 https://www.indiatoday.in/education-today/news/story/over-550-engineering-colleges-affiliated-with-aicte-shut-shop-in-last-6-years-1855266-2021-09-21
5 https://timesofindia.indiatimes.com/city/chennai/aicte-no-nod-for-new-engineering-colleges-till-2024/articleshow/90529253.cms
6 https://indianexpress.com/article/education/100-new-medical-colleges-in-next-5-years-by-upgrading-district-hospitals-health-ministry-8267263/
7 The NITs recruit students through the JEE (Main) and are next most reputed after the older IITs. Only 5% of the total seats are reserved for female candidates in all institutes that are taking part in the seat allocation process through the JEE (Main).
8 https://www.indiaScienceandtechnology.gov.in/listingpage/women-engineering-colleges
9 https://manipal.edu/mu/admission/fees.html

References

Abraham, B. (2022). Another Student Dies by Suicide in Kota, Taking the Number of Deaths in 2022 to 15. *Indiatimes*, December 24, 2022. Available at: https://www.indiatimes.com/news/india/another-student-dies-by-suicide-in-kota-taking-the-number-of-deaths-in-2022-to-15-588453.html. Accessed on August 30, 2024.

AFMC Information Brochure. (2023). *Armed Forces Medical College, Pune*. Admission to MBBS course-2023. Available at: https://www.afmc.nic.in/admissions/Updated%20Information%20Brochure%20for%20MBBS%202023%20admission%20in%20AFMC%20Pune.pdf. Last accessed on August 30, 2024.

Agrawal, Meena J., Bedekar, Madhura Y., and Nair, Deepa S. (2015). Gender Differences in Motivational Factors Towards Medical Career Choice. *IOSR Journal of Research & Method in Education* (IOSR-JRME) 5(4): 53–56.

AICTE. (2012–2013). *Approval Process Handbook*. Available at: http://www.old.aicte-india.org/downloads/approval_process_12_13_051011.pdf Last accessed on November 13, 2024.

AICTE Statistics. (2024). *A Statistical Information of the Institutes in India. Program-Wise. Approved Intake Institute Type Wise.* Engineering and Technology. AICTE website. Available at: https://app.powerbi.com/view?r=eyJrIjoiZmJkZmMwNDYtNzJjYy00N2FlLThjMjAtZGFlN2Q1ODYyOTczIiwidCI6ImVkYjFiNjA2LWJlMzYtNDFlZi04ODRhLTJiMGMxNjY2OGU4NiJ9. Last accessed on August 30, 2024.

AISHE Report. (2010–2011). *All India Survey on Higher Education*. Government Of India, Ministry of Human Resource Development, Department of Higher Education, New Delhi, 2013. Available at: https://www.education.gov.in/sites/upload_files/mhrd/files/statistics/AISHE201011_0.pdf. Last accessed on August 30, 2024.

AISHE Report (2021-2022). *All India Survey on Higher Education.* Government Of India, Ministry of Education, Department of Higher Education, New Delhi. Available at: https://aishe.gov.in/aishe-final-report/. . Last accessed on August 30, 2024.

AISHE Report (2017-2018). *All India Survey on Higher Education.* Government Of India, Ministry of Education, Department of Higher Education, New Delhi. Available at: https://aishe.gov.in/aishe-final-report/. Last accessed on August 30, 2024

AISHE Report (2014-2015). *All India Survey on Higher Education.* Government Of India, Ministry of Education, Department of Higher Education, New Delhi. Available at: https://aishe.gov.in/aishe-final-report/. Last accessed on August 30, 2024

Anand, M. (2023). *CBSE Class 12th Result 2023 Declared: Girls Outshine Boys Once Again, 87.33% Students Pass*, StudyCafe.in, May 12, 2023. https://studycafe.in/cbse-class-12th-result-2023-declared-girls-outshine-boys-once-again-87-33-students-pass-220151.html. Last accessed on August 30, 2024.

Banerjee, R., and Muley, V.P. (2008). *Engineering Education in India, Draft Final Report.* Sponsored by Observer Research Foundation; Energy Systems Engineering, IIT Bombay Powai, Mumbai. Available at: https://www.academia.edu/28572868/ENGINEERING_EDUCATION_IN_INDIA. Last accessed on August 30, 2024

Bansal, S. (2024). Kota's Big Ads, Star Teachers Fail as Suicides Cause 40% Drop in Enrollments, *India Today,* August 27, 2024.https://www.indiatoday.in/education-today/news/story/kota-student-suicides-student-enrollment-drop-2587310-2024-08-27. Last accessed on November 13, 2024.

Basu, S., Zuo, X., Lou, C., Acharya, R., and Lundgren, R. (2017). Learning to Be Gendered: Gender Socialization in Early Adolescence Among Urban Poor in Delhi, India, and Shanghai, China. *Journal of Adolescent Health* 61(4S): S24–S29. Available at: https://www.ncbi.nlm.nih.gov/pubmed/28915988. Last accessed on August 30, 2024.

Bhattacharyya, R. (2023). Indian Women Queue Up in Droves to Pursue Higher Education in Foreign Destinations. *The Economic Times.* May 27, 2023. Available at: https://economictimes.indiatimes.com/nri/study/indian-women-queue-up-in-droves-to-pursue-higher-education-in-foreign-destinations/articleshow/100556875.cms?from=mdr. Last accessed on August 31, 2024.

Brandell, Gerd and Else-Marie Staberg. (2008). Mathematics: A Female, Male or Gender-Neutral Domain? A Study of Attitudes Among Students at Secondary Level. *Gender and Education* 20(5): 495–509. https://doi.org/10.1080/09540250701805771

British Council. (2014). *Postgraduate Student Mobility Trends to 2024. Education Intelligence.* October 2014. British Council. Available at: https://www.britishcouncil.org/sites/default/files/postgraduate_mobility_trends_2024-october-14.pdf. Last accessed on August 31, 2024.

Chakrabarty, B. (2010). Some Additional Aspects of Gender Asymmetry. *Current Science* 99(4): 420.

Chanana, K., ed. (1988). *Socialisation, Education and Women: Explorations in Gender Identity.* Orient Longman: New Delhi.

Chanana, K. (2001). *Interrogating Women's Education: Bounded Visions, Expanding Horizons.* Rawat Publications: Jaipur.

Dailybite. (2017). *Student Suicides Are Our Collective Shame.* Education Mustn't Be a Battle unto Death. 23-10-2017. DailyO, DailyBite, Variety. Available at: https://www.dailyo.in/variety/students-suicide-coaching-institutes-exam-pressure/story/1/20204.html. Last accessed on August 31, 2024.

Das, Upasak, and Singhal, Karan. (2017). Gender Differences In Mathematics Performance: Evidence from Rural India. Paper prepared for the IARIW-ICIER Conference New Delhi, India, November 23–25, 2017. Available at: http://old.iariw.org/India/singhal.pdf

Das, Upasak, and Singhal, Karan. (2023). Solving It Correctly: Prevalence And Persistence of Gender Gap in Basic Mathematics in Rural India. *International Journal of Educational Development* 96(2023): 102703. Available at: https://doi.org/10.1016/j.ijedudev.2022.102703. Last accessed on August 31, 2024.

Deb, Sibnath, Strodl, E., and Sun, Jiandong. (2015). Academic Stress, Parental Pressure, Anxiety and Mental Health among Indian High School Students. *International Journal of Psychology and Behavioral Sciences* 5(1): 26–34. https://doi.org/10.5923/j.ijpbs.20150501.04

Deccan Chronicle. (2017). Fewer Girls at Coaching Centres Too. *Deccan Chronicle.* Available at: https://www.deccanchronicle.com/nation/current-affairs/030717/fewer-girls-at-coaching-centres-too.html. Last accessed on August 31, 2024.

EY-FICCI. (2014). *Higher Education in India: Moving Towards Global Relevance and Competitiveness.* FICCI Higher Education Summit 2014. https://smartnet.niua.org/sites/default/files/resources/EY-higher-education-in-india_0.pdf. Last accessed on August 31, 2024.

Gafoor, K. Abdul. (2017). *Influence of Out-of-School Experiences and Learning Styles on Interest in Biology, Chemistry and Physics among Higher Secondary Boys and Girls in Kerala.* Paper presented at the National Conference on Quality Education in Present Educational Scenario, Umiam, Meghalaya, India, February 23–24, 2017. Available at: https://eric.ed.gov/?id=ED581524. Last accessed on August 31, 2024.

Gautam, Meenakshi. (2015). Gender, Subject Choice and Higher Education in India: Exploring 'Choices' and 'Constraints' of Women Students. *Contemporary Education Dialogue* 12(1): 31–58.

Gupta, Namrata. (2007). Indian Women in Doctoral Education in Science and Engineering: A Study of Informal Milieu at the Reputed Indian Institutes of Technology. *Science Technology and Human Values* 32(5): 507–533.

Gupta, Namrata. (2012). Women Undergraduates in Engineering Education in India: A Study of Growing Women's Participation. *Gender, Technology and Development* 16(2): 153–176.

Gupta, Namrata. (2015). Rethinking The Relationship Between Gender and Technology: A Study of The Indian Example. *Work Employment & Society* 29(4): 661–672.

Hercog, Metka, and van de Laar, Mindel. (2017). Motivations and Constraints of Moving Abroad for Indian Students. *Journal of International Migration & Integration* 18(3): 749–770.

INSA. (2004). *Science Career for Indian Women: An Examination of Indian Women's Access to And Retention in Scientific Careers.* Report. Indian National Science Academy: New Delhi. Available at: https://www.ias.ac.in/public/Resources/Initiatives/Women_in_Science/report.pdf. Last accessed on August 31, 2024.

JEE-Advanced Report 2023. *JEE Reports.* IIT: Madras. 2024. Available at: https://jeeadv.ac.in/reports.html. Last accessed on August 31, 2024.

JEE-Advanced Report 2016. Joint Implementation Committee Report. Vol 1. IIT Guwahati. August 2016. https://jeeadv.ac.in/reports/2016.pdf. Last accessed on November 13, 2024.

Kapadia, Shagufta. (2017). *Adolescence in Urban India: Cultural Construction in a Society in Transition.* Springer: India.

Krishna, V.V. (2001). Reflections on the Changing Status of Academic Science in India. *International Social Science Journal* 53(168): 231–246.

Larsen, M. (2011). *Vulnerable Daughters in India: Culture, Development and Changing Contexts.* Routledge: New Delhi.
Mathew, L. (2022). The Merit of Medicine: Science Aspirations in India. *Cultural Studies of Science Education* 17(2022): 701–726.
Miller, D.I., and Wai, J. (2015). The Bachelor's to PhD STEM Pipeline No Longer Leaks More Women Than Men: A 30-Year Analysis, *Frontiers in Psychology*, 6, Article ID 37. https://doi.org/10.3389/fpsyg.2015.00037. Last accessed on August 31, 2024.
Mittal, Richa. (2013). Gender Disparity in Engineering Colleges: Is the Difference in the Preferred Learning Style the Reason? *Institute for Learning Styles Journal* 1(Spring): 11–34. Available at: https://pdfs.semanticscholar.org/acb5/33b11568e9098c8f92263a4bd1f7c3aae70a.pdf. Last accessed on August 31, 2024.
Mukhopadhyay, Carol C. (1994). Family Structure and Women's Participation in Science and Engineering. In Mukhopadhyay Carol C. and S. Seymour (eds), *Women, Education and Family Structure in India,* Westview Press: Boulder, CO, 103–132.
Mukhopadhyay, Carol C. (2009). How Exportable are Western Theories of Gendered Science? A Cautionary Word. In Neelam Kumar (ed), *Women and Science in India: A Reader,* Oxford University Press: India, 137–177.
Nagarajan, R. (2016). More Women Study Medicine, but Few Practice, *The Times of India*, January 11, 2016. New Delhi. Available at: https://timesofindia.indiatimes.com/india/More-women-study-medicine-but-few-practise/articleshow/50525799.cms. Last accessed on August 31, 2024.
NAS. (2018). *National Achievement Survey. National Report Card. Learning achievements of Students, Class X (Cycle 2).* Department of School Education and Literacy, Ministry of Education, Government of India. NCERT 2021. Available at: https://nas.gov.in/report-card/2017. Last accessed on August 31, 2024.
NASSCOM. (2022). *India's Tech Industry Talent: Demand Supply Analysis.* February 2022. Community by NASSCOM Insights. Available at: https://community.nasscom.in/communities/emerging-tech/indias-tech-industry-talent-demand-supply-analysis. Last accessed on August 31, 2024.
NCEE. (2005). *India Education Report.* National Centre on Education and the Economy. http://www.ncee.org/wp-content/uploads/2013/10/India-Education-Report.pdf. Last accessed on August 31, 2024.
NCRB. (2021). Suicides in India. Chapter 2. In *Accidental Deaths & Suicides in India* 2021. National Crime Records Bureau. Available at: https://ruralindiaonline.org/en/library/resource/accidental-deaths-suicides-in-india-2021/. Last accessed on August 31, 2024.
NER. (2019). *National Employability Report for Engineers. Examining the Employability of Indian Engineers.* Research Report. SHL. Available at: https://www.scribd.com/document/653368364/National-Employability-Report-Engineers-en-In. Last accessed on August 31, 2024.
NIRF. (2024). *National Institutional Ranking Framework.* MOE. Available at: https://www.nirfindia.org/Rankings/2024/Ranking.html. Accessed on August 31, 2024.
NSF. (2017). *Women, Minorities, and Persons with Disabilities in Science and Engineering.* National Center for Science and Engineering Statistics Directorate for Social, Behavioral and Economic Sciences. National Science Foundation. https://wayback.archive-it.org/5902/20240719205600/https://www.nsf.gov/statistics/2017/nsf17310/static/downloads/nsf17310-digest.pdf. Accessed on August 31, 2024.

NSF (2023). Report. Science and Engineering Degrees Earned. S&E degrees awarded to women, by field and degree level: 2011 and 2020. Figure 7-3. Available at: https://ncses.nsf.gov/pubs/nsf23315/report/science-and-engineering-degrees-earned#overall-s-e-degrees-earned-by-underrepresented-minorities

Open Doors (2024). Open Doors 2024 Annual Data. November 18, 2024. US Department of State and IIE. Available at: https://opendoorsdata.org/annual-release/international-students/. Last accessed on November 19, 2024.

Orberg, J.W. (2017). Uncomfortable Encounters between Elite And "Shadow Education" in India—Indian Institutes of Technology and the Joint Entrance Examination Coaching Industry. *Higher Education* 76(1): 129–144.

Pallavi, S.K., and Rajkumar, G.C. (2011). Professional Practice Among Woman Dentist. *Journal of International Society of Preventive and Community Dentistry* 1(1): 14–19. https://doi.org/10.4103/2231-0762.86376

Parikh, P.P., and Sukhatme, S.P. (2004). *Women Engineers in India. Economic and Political Weekly* 39(2): 193–201.

Poonam, S. (2016). Why 57 Young Students Have Taken Their Lives in Kota. By Snigdha Poonam. *Huffpost.* Available at: https://www.huffpost.com/archive/in/entry/why-57-young-students-have-taken-their-lives-in-kota_n_10232456. Accessed on August 31, 2024.

Radhakrishnan, S. (2011). *Appropriately Indian: Gender and Culture in a New Transnational Class.* Orient Blackswan: New Delhi.

Rashmi. (2005). *Doctor or Engineer? Which Has More Money and Demand*? RediffGetahead. February 7, 2005. Available at: http://www.rediff.com/getahead/2005/feb/07rashmi.htm/. Accessed on August 31, 2024.

Sancho, David. (2015). *Youth, Class and Education in Urban India: The Year that Can Break or Make You,* Taylor Francis Ltd: Milton Park.

Sengupta, S. (2023). Kota Suicides: Panel Suggests Preventive Measures to De-Stress Students. *Hindustan Times*, August 29, 2023. Available at: https://www.hindustantimes.com/india-news/kota-suicides-panel-suggests-preventive-measures-to-de-stress-students-101693279971063.html. Accessed on August 31, 2024.

Singh, Vijay A., and Pathak, P. (2010). Gender Asymmetry in Selection Tests at the Pre-college Level. *Current Science* 98(11): 1432–1433.

Smart, P. (2023). With a Push in Female Supernumerary Seats, Number of Women Students Increases in IITs. *The Indian Express*, February 5, 2023. Available at: With a push in female supernumerary seats, number of women students increases in IITs (indianexpress.com). Accessed on August 31, 2024.

Solanki, A., and Kashyap, S. (2014). Medical education in India: Current Challenges and the Way Forward. *Medical Teacher* 36(12): 1–5.

Sondhi, Gunjan. (2015). Indian International Students: A Gender Perspective. In S.I. Rajan (ed), *India Migration Report 2015: Gender and Migration*, Routledge: New Delhi, 104–119. Accessed at: https://www.researchgate.net/publication/283018927_Indian_international_students_A_gender_perspective. Accessed on August 31, 2024.

Sood, Mamta, and Chadda, R.K. (2010). Women in Medicine. *Indian Journal of Gender Studies* 17(2): 277–285.

Thakkar, D., Sambasivan, N., Yardi, P., Sudarshan, P., and Toyama, K. (2018). *The Unexpected Entry and Exodus of Women in CS and HCI in India, CHI 2018*. Proceedings of the 2018 CHI Conference on Human Factors in Computing Systems. Paper No. 352, 1–12. Available at: https://dl.acm.org/doi/10.1145/3173574.3173926. Accessed on August 31, 2024.

UNESCO-UIS. (2023). *Global Flow of Tertiary Level Students. India.* UNESCO Institute for Statistics. Available at: https://uis.unesco.org/en/uis-student-flow#slideoutmenu. Accessed on August 31, 2024.

Varma, R. (2010). Computing Self-Efficiency among Women in India. *Journal of Women and Minorities in Science and Engineering* 16(3): 257–274.

Varma, R. (2011). Indian Women and Mathematics for Computer Science. *IEEE Technology and Society Magazine* 30(1): 39–46.

2

GENDER AND ENGINEERING EDUCATION

Introduction

Gender biases in engineering in India differ from the West. In the West, computer engineering, for instance, is associated with the masculine attributes of geekiness or anti-social hours, while in India, computer-related branches are considered women-friendly. However, older branches of engineering such as civil and mechanical are perceived as masculine. Differences in the popularity of the branches and in the reputation of the engineering institutes create a hierarchy among students. Engineering institutes are male-dominated in numbers and masculine in culture. Female students of the lowest caste constitute a minority within the minority. This chapter discusses how the various hierarchies of branch, institute, gender, and caste manifest themselves.

Engineering sub-domains

Within 'engineering and technology' there are more than 35 sub-streams listed in the government's All India Survey of Higher Education Report (AISHE Reports 2021–2022), with the highest undergraduate enrollment in computer engineering (or computer science and engineering, CSE in the IITs/ engineering colleges), followed by electronics (ECE), mechanical (ME), civil (CE), and electrical (EE) in descending order. Unlike many universities in the world, the educational system in India does not allow switching out of the engineering program while pursuing the undergraduation (UG) program. Hence, there is little possibility of attrition and there can only be a drop-out. The drop-out rate at the UG level is lower than at the PG level due to students (in PG) leaving in between for lucrative jobs. The drop-out rate in the

DOI: 10.4324/9781003596660-3

TABLE 2.1 Proportion of women to total undergraduate enrollment and degree awarded in engineering branches (2021–2022)

Branch rank as per total enrollment or degrees awarded	*Engineering branch*	*Women enrolled (%)*	*Degrees awarded (%)*
1	Computer*	34.2	38.5
2	Electronics	38.3	23.5
3	Mechanical	7.6	6.4
4	Civil	22.9	23.5
5	Electrical	26.0	28.6
6	IT and Computer^	37.6	42.0

Source: AISHE 2021–2022, Tables 12 and 35; ^listed separately and not in 'engineering and technology' in AISHE; *computer engineering is the same as computer science and engineering

UG program could be as low as 0.004% on an average, per year (Registrar IITK 2016). Of the 15 students that dropped out between 2011–2015 at IIT Kanpur, there was not a single woman.[1]

The women's proportion is higher in the branches of computer and electronic engineering than in other branches (Table 2.1). Although mechanical is the third most popular branch, the share of women in total enrollment in mechanical is a mere 7.6%. The proportion of degrees obtained by women in IT and computer applications at the UG level is 42.0% (AISHE Reports 2021–2022: Table 35). Thus, overall, women's preferred branch in engineering is related to computers. This is in contrast to some of the Western countries; in the US, the proportion of women earning degrees in computer sciences has been declining since 1995 and was 21% in 2020 (NSF 2023: 6). According to a UNESCO report (Huyer 2015: 93), computer science shows a steady decrease in female graduates since the year 2000, and this decline is particularly marked in the high-income countries. Between 2000 and 2012, the share of women graduates in computer science slipped in Australia, New Zealand, the Republic of Korea, and the US (Huyer 2015).

In most countries of the West, computers are associated with male attributes such as highly technical skills and a requirement of long and anti-social hours of work on the machine; and those working on it are perceived as 'nerds' or geeks, who are typically male (Ensmenger 2010; Wajcman 1991). According to Rasmussen and Håpnes (1991), the primary reason for the decline in the number of women entering computer science in Norway is the disliked 'nerd' image. In India, the preference of women for computer-related branches of engineering indicates that theories that associate computers with masculinity are not universally applicable.

In many Asian countries, including India, computers are not viewed as masculine. Large numbers of women are enrolled in Malaysia in computer science as the discourse is not perceived to be masculine (Lagesen 2008).

However, Lagesen's study (2008) shows that in Malaysia, computer science is distinguished from engineering, with the latter perceived as 'masculine'. In India, computer science and engineering (which is an engineering branch) does not have a stereotypical anti-social image (Saxena 2021). Computers (whether as science or engineering) are viewed as woman-friendly, and this is due to gendered assumptions. For instance, the jobs associated with computers are considered 'safe' for women, as they are office-based and require only mental and not physical strength (Varma 2010). On the other hand, social prejudices against women working with machines other than computers, and the masculine work culture at machine-based industries, account for the gender gap in mechanical engineering (ME). Daughters are usually discouraged from opting for ME by their families. Although ME is viewed as masculine in various countries, the proportion of women opting for ME is higher in developed countries. In the US, the proportion of women among the UGs in ME is about 13%; however, at the elite Massachusetts Institute of Technology, through the proactive approach of the department, it has reached 49.5% (O'Leary 2017). In the UK, women make up 10.8% of first-degree entrants in ME in 2018–2019 (Engineering UK Report 2020). The gendered view of engineering branches and the impact of elite/non-elite engineering institutes are discussed in the next section.

Gender through branch and institute differentials

A short exploratory study was conducted by the author to understand the differences in the impact of branches and institutes on men and women. The participants (male and female) in the study had obtained Bachelor degrees in Engineering from non-IIT colleges, and were either pursuing their Masters from an IIT or had recently completed their Masters from an IIT. These students belonged either to civil engineering (CE) or computer science and engineering (CSE). The participants were found on LinkedIn at a time when the Covid restrictions in the institutes were still in place. In-depth interviews were conducted online with nine such engineers between January to May 2022.[2] The questions were related to the role of family, educational, and career choices, placement, internship/industrial experience, and 'image' of Engineering. The interviews lasted for about 45 minutes, on an average. The interviews were recorded, after the assent of the participants, purely as voice recordings to avoid hesitation to video recordings from participants. Interviews were also written down manually in case the recordings were erased inadvertently. The interviews were followed-up through messages or emails when the responses lacked clarity. All the participants were from urban backgrounds, were in their mid-20s, and were unmarried. Pseudonyms have been used to protect the identity of the participants.

Participants in the study had opted for engineering not because of their passion for technical work, but because of social pressure, status, job opportunities, or lack of options other than engineering. An engineering degree offered them job security. This is verified by the fact that the engineering sector has a higher rate of hiring (31% of engineering graduates were hired in 2020) compared to other fields of study in the country.[3]

However, there is stratification among engineers owing mainly to the stratification among the institutes and among the branches. According to the participants, those who studied from local (state or private) colleges are deemed to be engineers with little value in the labor market. For instance, one participant (Anita, female, CE) said:

"If you are not from a reputed college, you might not be considered good (by society); you are not treated well. But if you are, they see you as 'haua' (enigma)...I have seen changes; I was in private and (then) in IIT... It matters in job, family; you get more importance".

This reflects on the lack of quality education in most local colleges so that their graduates are unemployable. IIT is coveted by the students from such colleges, to create a greater professional and social value for themselves. A well-reputed college implies high-profile jobs and salaries. In society, 'you get respect as an engineer only when they hear about your package' (Tashi, female, CE).

Engineering education has become placement (i.e., on-campus recruitment by companies) driven, which has stratified engineering branches. As the chairman, All India Council for Technical Education, reportedly said: 'Job availability is the basis of measuring "popularity" of any particular branch of engineering'.[4] A higher proportion of students from CSE find placements compared to CE, which is at the bottom of popularity. According to a report,[5] the IITs together placed only 57% of CE students, while overall 77% of the students (all branches) were placed in 2020–2021. In regional colleges, the situation for hiring civil engineers is worse; for instance, in an engineering college in Chennai, among those registered for placements, only 9% of the students from the CE were placed compared to 83% in the CSE. Further, according to the same report, CE attracts lower salaries and the difference between the average annual salaries of students of CE and CSE could be as high as 1,600,000 INR.

Due to this stratification, those who obtain a higher ranking in entrance exams opt for CSE, and the rest have to opt for branches lower in popularity. Hence, students are usually compelled to take up branches that might not be of their interest. Secondly, those in CSE might have a feeling of being intellectually superior to those in other branches. For instance, one participant (Tashi, female, CE) remarked:

"It feels from within ...no one says it — that if you have [an] interest in CS, you are unable to get it because of [a] low rank. We have forgotten what we like ..."

Although participants were aware that their branch of engineering is unique technically, they believed that recruiting companies gauge the value of an engineer usually by their software skills. Hence, even those in non-CSE branches learn coding.

The participants believed that while engineering by itself is not considered by the society as unsuitable for women, certain branches of engineering are viewed as inappropriate for women because jobs in those sectors are unsuitable for them. For instance, Shachi (female, CSE) said:

> ... society thinks that men are better in CE/ME, because you have to work on the plant, there are security issues; the staff is mostly male and they are not respectful to a lady.

Similarly, Hari (Male, CE) said:

> Yes, in CE (men are more suitable). If you go for [a] road inspection, you are staying elsewhere, need separate rooms; (there is) lot of physical work, dust in fieldwork where mostly male [sic] are working there.

The contrast in the social perception regarding CE and CSE is brought out in Kirti (CSE, female): 'Decision to join CSE was mostly a method of elimination because CE/ME are mostly male preferred. And IT sector is booming, there is no crisis of job; parents also thought that CS is a good option from girls' perspective.'

Engineering education in local engineering colleges is not oriented towards the market (NER 2016). This is true for all branches, but it has a substantial impact on those branches requiring practical training in the field/industry, which in turn affects women more than men during placements. Internships or practical training is a requirement for a bachelor's degree. However, internships are not emphasized in many colleges, and the students are often able to manipulate the system to get training certificates. Further, 'what is taught (in the local colleges in BTech) is old, and not useful ...The skill wanted by industry, is not given' (Hari, CE, Male). Women in CE tend to opt for desk work in internships, avoiding fieldwork. As Nalin (Male, CE) said, 'In internship, I didn't see any female intern, but heard ..., that women will not go to the fieldwork ... maybe because of harassment. Women in my branch in XXX (UG college) went for design jobs, which are desk jobs; which kind of supports what I heard.' Conditions on the site during CE internships are gendered, as noted by Anita (female, CE):

> there is an uncomfortable feeling in general because you are one female in internship ... it's very hard to make others (workers on the construction site, to learn what's going on) open up or speak freely compared to a male counterpart.

Apprehensions regarding field internships discourage parents, as in the case of Meenal (female, CE): 'Environment in field work is not good. Family will not allow (for internships) as there are mostly males at work(place)'.

A lack of field internship and the resulting field inexperience constructs the deficit in 'STEM capital' (Cohen et al. 2021), affecting the initial placements in CE among women. It appears that the core companies are less willing to hire women for the initial jobs after a bachelor's or master's. Further, women themselves are wary of field jobs. For instance, Meenal (female, CE): 'Reason why women in CE do not have many job opportunities is because they lack experience in this field and also, they do not want to work in such jobs'. Negative experiences during field internships discourage women from opting for construction jobs. Thus, Anita (female, CE), because of her experience in an internship in bachelor's, was no longer excited about the construction sector and, therefore, thought of pursuing higher studies (i.e., master's).

The situation in CSE is quite different. Internship in CSE does not require field work. The projects done during a bachelor's course work are sufficient for placements. As Shiv (CSE, Male) said:

> I have not done any industrial project. But as a CSE student, it's great. You are building programs ... so many problems are there, we can find solutions ... It is so powerful. Even companies that hire (from) other branches of engineering, want programming. During the placements, the companies opted for software engineering. There weren't many jobs for ME/CE. Everything is digitalized now. I am satisfied.

Women in CSE might be *preferred* by companies to maintain a diversity ratio. For instance, Kirti from CSE said: 'In my experience with placements, I found that companies like to maintain 50-50 male/female ratio, and so, prefer female sometimes.' Rohan (male) from CSE had a similar opinion: 'I do feel there is a gender bias, as far as career opportunities are concerned. Most corporations prefer to employ female candidates. ...'

It is not surprising, therefore, that women represent 51% of entry level hiring in the IT sector (NASSCOM 2016) and the highest proportion of about one-third of the workforce in this sector (CFA 2023). Similar figures are not available for CE. However, while there are 22% of women engineers in IT service companies, women constitute about 7.7% to 10% of the workforce in many core industries (NER 2016). According to one estimate, women constitute only about 1.4% of the technical/managerial personnel (including architects, civil engineers, supervisors) in the construction sector (PP-WTC report 2023).

Computer science and engineering is a unique example where a high reputation of the field coincides with the field's perception as friendly for women. Yet, despite being perceived as woman-friendly, there is not an overflow of women in the field of CSE. In fact, the growth of women's participation in CSE education is much lower (25%) than that of men (105%) between 2015 and 2021, as noted in the previous chapter. The growth in CSE could be due to the emergence of new technologies such as Artificial Intelligence (AI), Data Science, Block-chain Technology, and Machine Learning. The lower growth of women participating in CSE does not augur well for gender equality. That the new technologies have become gendered is indicated by the low proportion of merely 22% of women professionals in AI in India (UNESCO Science Report 2021), with gender gap in GenAI ranging from 35% in junior positions to as high as 63% in senior management positions (Jain 2024). There could be two reasons for this: one, bias in the labor market so that women are less likely to be hired for AI roles than men as found elsewhere (Young, Wajcman, and Sprejer 2023); and two, women's upskilling is lagging behind those of men's. This could be due to gendered reasons, such as women lacking time after office hours, particularly when married and/or the inability to shift companies and to be mobile. Research is lacking on this aspect. However, recent media reports indicate an upsurge in women upskilling themselves in AI courses in the June 2024 quarter (Ghai 2024). Whether this corresponds with greater AI roles for women in the industry is yet to be seen.

Undergraduate women outperform in gendered social worlds

Studies conducted (Gupta 2020; ICSSR Report 2017; Prakash and Yadav 2024) and an IIT-Delhi survey of their 2012 entry batch[6] have found that female undergraduates outperform their male counterparts in elite engineering institutes (Table 2.2). Further, a study conducted between an IIT and a private university (ICSSR Report 2017) showed that there was no difference in the scores on self-efficacy (using Schwarzer and Jerusalem's scale) among men and women undergraduates. This is in contrast to the Western literature that indicates self-efficacy beliefs as a major factor in gender gaps in participation in computer science, engineering, and physics (Cheryan et al. 2017).

In the formal class environment in the IIT-X (Gupta 2020), no difference was found between men and women regarding asking questions in class to clarify doubts. Also, women believed more than their male counterparts that faculty encouraged class interaction (Gupta 2020). However, in the study of a private university (pseudonym as PRC), the perception of students believing

TABLE 2.2 Mean CPI (out of 10) of men and women in BTech at IIT-X and PRC

Institute	*Gender*	*N*	*Mean*	*Std. deviation*	*Std. error mean*	*Sig. (p)*
PRC	Women	228	7.76	1.26	.084	.00**
	Men	460	7.22	1.32	.062	
IIT-X	Women	151	7.53	1.4	.113	.04*
	Men	515	7.27	1.4	.061	

***p<.01; *p<.05 Source:* ICSSR Report 2017

that faculty was biased in favor of boys was significantly high among both girls and boys; and, in the male dominated branches such as CE, women felt that teachers were sexist and might make such types of remarks in the classroom (ICSSR Report 2017). This suggests that in private institutes, a formal environment might be more gender-biased than in the IITs.

Women, being in the minority in the engineering institutes, confront 'tokenism' in the informal environment. Kanter (1977) identified 'tokenism' as the effect of low numbers of women in male-dominated organizations that includes performance pressures, social isolation, and role encapsulation for women. However, gender relations, and not just the numerical proportion of women, are significant in the occurrence of tokenism (Yoder 1991).

As Table 2.2 shows, comparing IIT and PRC, women outperformed men in both institutes. This seems to verify Sonnert and Fox (2012) that there is a positive relationship between 'tokenism' and academic performance among undergraduate students; *that women tend to overachieve when men predominate in order to counter their minority status.*

The minority status manifests in various ways. Since women in most courses are few in numbers, they lack female colleagues from whom to seek help for academics. Stereotypes that women are less suited for technical tasks tend to operate in teamwork; for instance, women in a team might be seen as more suitable for drawing or compiling a report rather than casting, welding, or cutting (Gupta 2020). There is a greater culture among men with respect to computer games and learning computer languages as group activity that facilitates their learning. Thus, one-third of women participants at IIT-X believed men have a greater ability to work with computers, while only one-tenth thought that men have superior mathematical and problem-solving skills. The lack of social capital forces women to rely on their own merit (human capital), thus, having to work harder than boys (Mittal 2011).

The informal interaction among engineering students reflects conflicting desires among men and women, resulting from the patriarchal cultural context. For men, there is a desire to be free from parental authority, and consequently, indulging in irreverences (Bansal 2013); for women, there is a desire to be independent and yet maintain a commitment to family and morality (Thapan 2007). Further, with dating being uncommon, particularly in the pre-college period, coupled with an intense study required in the last two years of school (before joining the engineering college), students are deprived of interaction with members of the opposite sex (who are not related). According to Ria (female, EE, 7th semester):

> (boys) … are too desperate to have girlfriends. Having studied for JEE for two years, they have not had any interaction with girls. College time is visualized as a time for such interaction, influenced by movies and all …
> *(Gupta 2020: 12).*

Men are considerably more involved in non-academic activities, which is partly a reflection of the freedom from parental authority. Compared to women, men are also under greater peer pressure to participate in such activities, including outdoor sports, cultural, and technical. Women lack leadership positions in student bodies. At IIT-X, members in the Students' Senate, nominees of the Student Gymkhana, heads of organizing committees for student festivals were usually men (Gupta 2020). Further,

> All meetings (of student bodies) are usually held after midnight and this is not suitable to many of us; also, boys (tend to) smoke, drink … don't want girls around, use loose language amongst themselves.
> *(Seema, MSE, 6th semester).*

Gendered positions in student bodies impart power to men in the peer group and thereby contribute to a hegemony of men and men's practices. There are instances of sexist humor through innocuous jokes, particularly in groups or through the internet (Mittal 2011; Gupta 2020). However, some men dislike this aggressive masculinity, and many of the male students behave differently when they are not in a group (Gupta 2020).

In PRC (ICSSR Report 2017) too, where women in bachelor's programs accounted for 20% of the undergraduate engineering students, problems related to tokenism manifested themselves. Men formed their own informal groups and women usually did not form a part of such groups. This created an information deficit among women due to which they had to struggle more in working out projects, dealing with changes in class schedules, etc. and were often sidelined for important tasks.

Most hostels, whether in government or private engineering institutes (apart from some IITs), have different timings for hostel entry of men and women due to 'security' reasons, and women are recalled to their hostels in the evenings earlier compared to boys. Additionally, many colleges implement the rules more strictly for women than for men. Both the early recall of women and strict implementation have gendered implications. For instance, at PRC (ICSSR Report 2017), the deadline for women to return to the hostel was 8:30 or 8:00 p.m. (summer and winter timings respectively) and for men, 10:30 p.m. While this is discriminatory enough, the men's timings were not strictly adhered to. The timings translated into denial of access to resources and created several other problems for women. It affected access to the library or any help that they might need from their male/female colleagues (in other hostels) after 8:30 p.m. at night. The earlier deadline for women restricted access to leadership positions or acting as an organizer of events because it required attending meetings and working after 8:30 p.m. Usually one rises to the headship of societies through hard work and skills. However, headship for women was a problem due to the 8:30 pm rule as late-night meetings would be difficult. It also created problems in collaborating on projects. This suggests the existence of gendered policies and systems in institutes, that are seemingly rational and gender neutral.

However, there has been a greater systemic response to sexual harassment of women in institutes since the 2013 framing of the rules in this regard. Internal complaints cells in the institutes, made mandatory by the government, have been set up in educational institutes.

The presence of women who passed a tough exam (for instance, the JEE) is in itself a challenge to the masculine image of engineering. Male students recognize that some women might be equally or more intelligent and perform better academically than them. Most women are critical of the traditional norms that inhibit aspirations of women, and they have high expectations and confidence regarding their career advancement (Gupta 2020).

At single-sex (women's) engineering colleges, 'tokenism' is absent; however, there are gender-related issues. As one study reported (Gupta 2012), concern for women's security led to a close, almost repressive supervision. There are several rules and regulations that restrict students' movement off campus. For day scholars, there are special college buses operating in the city; as one student said,

> If a day scholar wants to go back home after coming to college because she is unwell, the basic assumption is that she is pretending, and long-winded procedures have to be followed, so the whole purpose is lost.
>
> *(Gupta 2012: 169).*

According to the director of the institute, parents expect and demand strict security arrangements from the authorities.

Gendered roles and an engineering degree

The parental role in not only educational but also marital decisions reproduces a gender hierarchy in engineering education. For the women that graduate from the IITs, high grades are important not only because they provide the sole probability of procuring a good job in a male-dominated market, but also because there is socio-cultural pressure to perform. A good academic performance is desirable to avoid being coerced into getting married after BTech. According to Shruti (female, CSE, 5th semester, BTech, IIT-X):

> For women, it is more like perform or get married', implying that if women perform well, they would be 'allowed' by the family to pursue their career further.
>
> *(Gupta 2020: 11).*

While men are expected to grow themselves in their careers, women face pressures to get married due to the age hypergamy in society (women marrying at a younger age than men). This again reflects the patriarchal considerations of a 'public-private' dichotomy of middle class, preferring their daughters to be independent financially rather than aiming for highly successful careers.

For men and women, engineering degrees (BTech) from local/regional institutes might not fetch placements. The struggle for jobs among the non-CSE engineers who have bachelor's from a local or non-elite college hurts women more than men. In the study on master's students (referred to in the section 'Gender through branch and institute differentials'), Tashi referred to the plight of such graduates (female, CE):

> There are so many people who are not getting a decent job. Situation is really bad for girls [aged] between 25 and 30 years, because at this age parents pressurize [daughters] to get married. Many girls from my college [of undergraduation] have got married. They lose their choice. If they had got a job, they would have decided when to get married. If you don't get a job and you are staying with family, you work in the kitchen, the male of the family says why not get her married. At this point, when you are not getting a job, relatives also pressurize. If you have a job, you are staying away from your family. They cannot force you much. This is a harsh reality.

Tashi's statement suggests that woman engineers that fail to get job could be pressurized to marry, which implies aborting a career in engineering. In this study, female participants in CE and CSE perceived age-based marital pressures, as for instance, Kirti (female, CSE):

> I don't have martial pressure for now but yes, the hard deadline (given by family) of getting married by 27–28 [years of age] sometimes bothers me.

or Meenal (female, CE), who does not face any pressure to get married now, but said 'will get it after MTech'.

Intersectionality of gender and caste in an engineering education

India adopted positive discrimination measures to promote social equality, initially for the SC/STs, and later the OBC-NCLs (the economically backward among the OBCs or the non-creamy layer). Since caste and class roughly correspond, the SC/STs are at the bottom of the social, economic, and educational ladder (Bharti 2018). The benefits of affirmative action have improved the educational attainment of marginalized caste groups, but these gains have been concentrated among males with lesser effect for females (Cassan 2019). Females and individuals from historically disadvantaged castes are significantly less likely to study science (Kumar and Sahoo 2022). Girls from lower classes are not fearful of science and mathematics compared to boys; but there is a sexist behavior on the part of students and teachers. Discouragement from studying science and mathematics for these girls also arises from the common patriarchal factors, such as focus on the son's education due to his earning capacity and parental care obligations and the daughter's engagement with the household tasks (FAT STEM Report 2014).

Privatization has increased the cost of education, and demand for new skills (beyond formal education) is rising, affecting its access among the Dalits (SCs)/Adivasis (STs). The Dalit proportion in education is lower than their proportion in the population at all levels, and that of female Dalits falls even lower than their male counterparts. Thus, though the proportion of the Dalit females is about 16% of the female population in the country (97.8/586.4 million as per the 2011 Census), the Dalit women constitute only about 10.7% of women enrolled in BTech (AISHE Reports 2021-2022, Table 11). The Dalit women lag behind their male counterparts constituting only 29% of the total Dalit students enrolled in BTech (AISHE Reports 2021-2022, Table 11). On the other hand, in BSc, a less prestigious course with lower job prospects compared to engineering/medical degrees, has about 15% of Scheduled Caste women as the proportion of women enrolled, comparable to their proportion in population; and, female Dalits are half of the Dalit enrollment in the course (AISHE Reports 2021-2022, Table 11).

Studies on the relationship between education and caste have found preconceived stereotypes and implicit biases among teachers. For instance, there is discrimination in grades awarded to the lower castes due to lower expectations (Parashari 2019). Caste discrimination in academia could be structural, inadvertently constructed in policies and practices through caste-based categorizations; socially stigmatized as 'not capable of success' and as 'destined to fail', it affects the SC/ST students' academic achievement as well as social adjustment (Rao 2013). The caste bias is deep-rooted as exemplified in the casteist abuses hurled by a female upper caste professor in an IIT at her students, suggesting an intersectionality of caste and organizational position, that overshadows gender.[7]

Caste creeps into the interaction on the campuses. As Rao (2013) in his study of the IITs found, the Dalit/Adivasi students become visible right from the introductions in the class where teachers might ask name, rank, place of origin, etc.; that they are not welcome in the team work, lab, or workshop groups and even in the informal groups, unless they distinguish themselves. The notions of purity/pollution exist in interactions (e.g., an upper caste student's unwillingness to share a room), and the possibility of inter-caste romance is negligible.[8] The humiliation and isolation have led to suicides by Dalit students, accounting for half of the total suicides in the IITs, and the psychological support structures at the IITs are passive; their welfare boards unable to address their needs (Deep and Adarsh 2023). The Deep and Adarsh (2023) study finds a contrast between the theory and practice of inclusion at the IITs. IIT Bombay's recent guidelines, prohibiting students from asking each other's ranks or make casteist remarks (Lakshman 2023), officially affirm the prevalence of casteism in interactions on the campuses.

The isolation of the Dalit/Adivasi students is accompanied by low academic performance. The stereotype threat, or the fear that being low caste, would lead to low reward, adversely affects performance (Hoff and Pandey 2006). A study on the students of an IIT (Gupta 2019) found that the Cumulative Performance Index (CPI out of ten) of the Dalits/Adivasis is lowest among all caste groups. Further, women of Dalit/Adivasi groups have the lowest CPI compared to the women of other castes. However, similar to the pattern in other caste groups, women had a higher CPI than men within their own caste. A higher CPI among women, as noted earlier, is a product of 'tokenism'. Further, while women of all castes had higher levels of alienation compared to the men of their caste, reflecting gender effects, the intersection of caste and gender is evident in the finding that *Dalit/Adivasi women were the most alienated* (that is, socially isolated, lacking in self-worth and belongingness to the institute). They also scored lowest on self-efficacy (one's ability to perform successfully in a wide variety of achievement situations) among all castes, male and female groups.

Caste and gender work in the favor of the men of General Castes (or the upper castes). The study found that self-efficacy and self-esteem are the highest for General Caste male students among all caste groups, and male/female. The male students of the General Castes are at the top of the social ladder, reflected in terms of being the least alienated, having the highest faith in their mathematical and problem-solving abilities and highest self-esteem and self-efficacy among all caste and gender groups. The female students of

General Castes lead on parameters such as confidence in their abilities, self-efficacy, and CPI among women of all caste groups.

This reflects the role of 'cultural capital' in reproducing inequalities. Bourdieu (1977) emphasized the non-material resources (e.g., cultural goods, language, qualifications) possessed by higher classes' households which are passed on to their children. Although, referring to 'class' in Bourdieu, in India, caste and class correspond, and the lower castes fail to accumulate the 'cultural capital'. The education system demands 'linguistic and cultural competence and that relationship of familiarity with culture, which can only be produced by family upbringing when it transmits the dominant culture' (Bourdieu 1977: 494). A lack of 'cultural capital' among the lowest caste reproduces caste inequality in prestigious engineering and technology education. Women of the lowest caste are impacted by both gender and caste hierarchy.

Conclusions

Although engineering has been male-dominated and considered masculine historically, studies show a parental encouragement of engineering education for daughters in recent decades. However, it is viewed favorably for patriarchal reasons, and families might not be enthusiastic about the successful careers of their daughters. Further, proportions continue to be low in engineering education compared to other fields, particularly in the older branches of engineering such as CE/ME. The societal gender expectations of a son as the potential breadwinner, fuelled by the glamour of the engineering education from elite institutes and prestigious jobs subsequently, construct women, viewed central to domestic roles, as subsidiary to the domain.

Further, there are systemic issues that feed in to the societal stereotypes. The nature of institute, elite versus non-elite, affects job opportunities. Lack of jobs brings marital pressures on women, threatening their career. There are gender stereotypes against some of the branches of engineering. Recruiters tend to exhibit bias against women with regard to certain branches such as civil engineering during placements. Women from CSE might not be discriminated against at the recruitment stage, however, can we say the same for the leadership stage? This is discussed in the next chapter.

The engineering education system reproduces gender hierarchy in the non-CSE fields. For instance, internships in CE involve field experience and inefficiencies of the education system at the local colleges, a lack of rigorousness in internships, women's field work being tough due to conditions at the workplace, and experience with the placements combined with the stereotypes, makes CE women-unfriendly. On the other hand, CSE graduates have job readiness even if lacking in industrial experience, due to the projects done during their study and the knowledge of software.

Women in prestigious engineering educational institutes face consequences of an overlap of gender hierarchy and minority status, one of them being performance pressure. Informal interaction is marked by hegemony of men on the campuses, reflected in student bodies, cultural, and technical activities. In single-sex institutes, these issues are absent but women might be closely monitored, reflecting patriarchal norms.

Intersectionality of gender and caste is manifested in engineering colleges in various ways. The women engineering students of the lowest caste have the highest feeling of 'alienation' (that is, socially isolated, lacking in self-worth, and lack of a sense of belonging to the institute). Caste is significantly correlated to academic performance, and women of the lowest caste have the lowest performance among women.

Computer engineering is a rare case where the domain is highly reputed and yet it is perceived as friendly for women. However, the growth of women's enrollments in this field is much lower than that of men. Also, how does the women-friendly image of CSE reflect in the IT (information technology) organizations? Further, as noticed in the earlier chapter, there is a growing presence of women in STEM fields, particularly the rising proportions of women at the doctorate level in science as well as engineering, and in medicine at the master's level, suggesting heightened career aspirations and change. Does the proportion of women in STEM professions reflect trends similar to those witnessed in education? What is the nature of relationship between gendered expectations and women's professional roles? Does this relationship manifest changes in recent years? These and other related questions are dealt with in the next chapters.

Notes

1 List of withdrawn students from Academic Section, IIT Kanpur communicated on March 30, 2017 through e-mail.
2 4 CSE and 5 CE students; 5 were female; admitted in Masters' between 2019–2021.
3 "Rate of hiring amongst engineers across India from 2014 to 2020". Published by Manya Rathore, September 13, 2022 on the website of Statista, Statista Inc., US. Available at: https://www.statista.com/statistics/1043283/india-hiring-rate-engineers/#:~:text=Hiring%20rate%20for%20engineer%20graduates%20across%20India%202014%2D2020&text=The%20rate%20of%20hiring%20among,of%20study%20in%20the%20country
4 Popularity of engineering streams depends on job scenario: https://timesofindia.indiatimes.com/home/education/news/popularity-of-engineering-streams-depends-on-job-scenario/articleshow/90319832.cms
5 Based on the replies to Right to Information (RTI) applications filed by Careers360: https://engineering.careers360.com/articles/btech-civil-engineering-placements-branch-wise-salary-premium
6 Initiatives for gender equity and sensitization, IIT-Delhi: https://iges.iitd.ac.in/jeew/supernumerary-seats-women.pdf
7 IIT-Kharagpur professor abusing SC, ST students is not a one-off (qz.com).
8 Caste in IIT-Kanpur: https://voxiitk.com/caste-in-iit-kanpur-caste-based-social-realities-that-shape-our-academic-existence/

References

AISHE Reports (2021–2022). *All India Survey on Higher Education.* Government Of India, Ministry of Education, Department of Higher Education: New Delhi. Available at: https://aishe.gov.in/aishe-final-report/. Last accessed on September 1, 2024.

Bansal, Parul. (2013). *Youth in Contemporary India. Images of Identity and Social Change.* Springer: India.

Bharti, N.K. (2018). *Wealth Inequality, Class and Caste in India, 1961–2012.* World inequality database, world inequality lab. Paris School of Economics, WID. World Working Paper N°2018/14. Available at: https://wid.world/document/n-k-bharti-wealth-inequality-class-and-caste-in-india-1961-2012/. Last accessed on September 1, 2024.

Bourdieu, P. (1977). Cultural Reproduction and Social Reproduction. In L. Karabel and A.H. Halsey (eds), *Power and Ideology in Education*, OUP: Oxford.

Cassan, G. (2019). Affirmative Action, Education and Gender: Evidence from India. *Journal of Development Economics* 136(2019): 51–70.

CFA. (2023). *Mind the Gender Gap Analysis of Women's Participation, Pay, and Other Measures in Indian Public Companies.* By Sivananth Ramachandran, Jolly Balva, and Meera Siva. CFA Institute. Available at: https://www.cfainstitute.org/-/media/documents/article/industry-research/Mind-Gender-Gap-Indian-Public-Companies.pdf. Last accessed on September 1, 2024.

Cheryan, S., Ziegler, S.A., Montoya, A.K., and Jiang, L. (2017). Why Are Some STEM Fields More Gender Balanced Than Others? *Psychological Bulletin* 143(1): 1–35. https://doi.org/10.1037/bul0000052. Epub 2016 October 10. PMID: 27732018.

Cohen, S.M., Hazari, Z., Mahadeo, J., Sonnert, G., and Sadler, P.M. (2021). Examining the Effect of Early STEM Experiences as A Form of STEM Capital and Identity Capital on STEM Identity: A Gender Study. *Science Education* 105(6): 1126–1150.

Deep, P., and Adarsh, T. (2023). Multi-Layered Exclusion in Premier Higher Education Institutions of India. *Higher Education* (Published Online September 8, 2023). Available at: https://doi.org/10.1007/s10734-023-01102-z. Last accessed on September 1, 2024.

Engineering UK Report. (2020). *Engineering UK 2020. Educational Pathways into Engineering.* Available at: https://www.engineeringuk.com/media/232298/engineering-uk-report-2020.pdf. Last accessed on September 1, 2024.

Ensmenger, N. (2010). *The Computer Boys Take Over: Computers, Programmers, and the Politics of Technical Expertise.* MIT Press: Cambridge, MA.

FAT STEM Report. (2014). *Why Don't Girls opt for STEM (Science, Technology, Engineering, Maths)? A Gendered Enquiry.* Available at: https://www.fat-net.org/sites/default/files/resources/FAT_STEM_GIRLS_Report_2014.pdf. Last accessed on September 1, 2024.

Ghai, N. (2024). Women Professionals Are Upskilling Like Never Before as They Bet on Potential of AI, *The Economic Times,* July 14, 2024. Available at: https://economictimes.indiatimes.com/jobs/hr-policies-trends/women-raising-their-skills-game-in-drive-to-thrive/articleshow/111719776.cms?from=mdr. Last accessed on September 1, 2024.

Gupta, Namrata. (2012). Women Undergraduates in Engineering Education in India, A Study of Growing Women's Participation. *Gender, Technology and Development* 62(2): 153–176.

Gupta, Namrata. (2019). Intersectionality of Gender and Caste in Academic Performance, Quantitative Study of an Elite Indian Engineering Institute. *Gender, Technology and Development* 23(2): 165–186.

Gupta, Namrata. (2020) Patriarchy Reinforced or Challenged? A Study of Engineering Students in an Elite Indian Institute. *Gender, Technology and Development* 24(2): 250–270.

Hoff, K., and Pandey, P. (2006). Discrimination, Social Identity, And Durable Inequalities. *American, Economic Review* 96(2): 206–211.

Huyer, Sophia. (2015). Is The Gender Gap Narrowing in Science and Engineering? In *UNESCO Science Report: Towards 2030*. UNESCO, 2015, 85–103. . Available at: https://unesdoc.unesco.org/ark:/48223/pf0000235447. - . Last accessed on September 1, 2024.

ICSSR Report. (2017). *Women in Undergraduate Engineering in India: A Study of Gender in Academic Environment*. By Namrata Gupta. Report submitted to ICSSR (vide F.No. 02/248/2014-15/ICSSR/RPR). Available at: https://www.researchgate.net/profile/Namrata-Gupta-2/publications. Last accessed on September 1, 2024.

Jain, N. (2024). Ethical AI - Women's Critical Role in Shaping the Future of Technology, NASSCOM community, September 2, 2024. Available at: https://community.nasscom.in/communities/women-leading-change/ethical-ai-womens-critical-role-shaping-future-technology. Last accessed on November 13, 2024.

Kanter, R. (1977). *Men and Women of the Corporation*. Basic Books: New York.

Kumar, A., and Sahoo, S. (2022). Caste, Gender, and Intersectionality in Stream Choice: Evidence from Higher Secondary Education in India (September 26, 2022). Available at: https://ssrn.com/abstract=3894538. Last accessed on September 1, 2024.

Lagesen, V.A. (2008). A Cyberfeminist Utopia?: Perceptions of Gender and Computer Science among Malaysian Women Computer Science Students and Faculty. *Science, Technology, & Human Values* 33(1): 5–27.

Lakshman, A. (2023). IIT-Bombay Dalit Student Death. Senior Says Darshan Solanki Felt Alienated by Roommate, *The Hindu*, March 13, 2023. Available at: https://www.thehindu.com/news/national/iit-bombay-dalit-student-death-senior-says-darshan-solanki-felt-alienated-by-roommate/article66611752.ece. Last accessed on September 1, 2024.

Mittal, Parul A. (2011). *Heart Breaks and Dreams: Girls @ IIT*. Srishti Publishers and Distributors: New Delhi.

NASSCOM. (2016). *Women 'In'Equality-Not Anymore!: Gender Diversity & Inclusivity Trends In The IT-BPM Sector*. March 2016. Available at: https://nasscom.in/knowledge-center/publications/women-inequality-not-anymore-gender-diversity-inclusivity-trends-it. Last accessed on September 1, 2024.

NER. (2016). *National Employability Report 2016*: Engineers. Aspiring Minds. Annual Report 2016. Available at: https://quantumuniversity.edu.in/qlib/document/pdf/National%20Employability%20Report%20-%20Engineers%20Annual%20Report%202016.pdf. Last accessed on September 1, 2024.

NSF. (2023). Report. Science and Engineering Degrees Earned. National Center for Science and Engineering Statistics. NSF January 30, 2023. Available at: https://ncses.nsf.gov/pubs/nsf23315/report/science-and-engineering-degrees-earned. Last accessed on November 13, 2024.

O'Leary, Mary Beth. (2017). Closing The Gender Gap in Mechanical Engineering, *MIT News*, July 31, 2017. http://news.mit.edu/2017/closing-the-gender-gap-in-mit-mechanical-engineering-0731. Last accessed on September 1, 2024.

Parashari, S. (2019). *Teacher Discrimination in Occupational Expectations and Grading*. ISS Working Papers – General Series 640, International Institute of Social Studies of Erasmus University Rotterdam (ISS), The Hague. Available at: https://pure.eur.nl/en/publications/teacher-discrimination-in-occupational-expectations-and-grading. Last accessed on September 1, 2024.

PP-WTC report (2023). *Pink Collar Skilling: Unleashing the Women's Power in the Real Estate Sector*. Primus Partners. Available at: https://www.primuspartners.in/reports/pink-collar-skilling-unleashing-the-women-power-in-real-estate-sector. Last accessed on November 14, 2024.

Prakash, A., and Yadav, I.S. (2024). Disparity in Employment in India: Evidence from Engineering Graduates Across Social Groups and Gender. *Review of Social Economy*: 1–20.

Rao, S.S. (2013). Structural Exclusion in Everyday Institutional Life. Labelling of Stigmatized Groups in an IIT. In Geetha B. Nambissan and S. Srinivas Rao (eds), *Sociology of Education in India. Changing Contours and Emerging Concerns*, OUP: New Delhi, 199–223.

Rasmussen, B., and Håpnes, T. (1991). Excluding Women from The Technologies of The Future: A Case Study of the Culture of Computer Science. *Futures* 23(10): 1107–1119.

Registrar Office, IIT Kanpur. (2016). Information Regarding Provisionally Admitted Question for the Lok Sabha, 16 November. Kanpur: IIT Kanpur.

Saxena, P. (2021). Gender and Computer Science Debate at Indian Institutes of Technology. *International Journal of Gender, Science and Technology* 13(2): 88–109.

Sonnert, G., and Fox, M.F. (2012). Women, Men, and Academic Performance in Science and Engineering: The Gender Difference in Undergraduate Grade Point Averages. *The Journal of Higher Education* 83(January/February 2012): 73–101.

Thapan, Meenakshi. (2007). Adolescence, Embodiment and Gender Identity: Elite Women in a Changing Society. In Rehana Ghadially (ed), *Urban Women in Contemporary India: A Reader*, Sage: New Delhi, 31–45.

UNESCO Science Report (2021): To Be Smart, The Digital Revolution Will Need to Be Inclusive. Excerpt from the UNESCO Science Report. Available at: https://unesdoc.unesco.org/ark:/48223/pf0000375429. Last accessed on November 14, 2024.

Varma, R. (2010). Computing Self-Efficiency Among Women in India. *Journal of Women and Minorities in Science and Engineering* 16(3): 257–274.

Wajcman, Judy. (1991). *Feminism Confronts Technology*. Pennsylvania State University Press: Pennsylvania.

Yoder, J.D. (1991). Rethinking Tokenism: Looking Beyond Numbers. *Gender & Society* 5(2): 178–192.

Young, E., Wajcman, J., and Sprejer, L. (2023). Mind the Gender Gap: Inequalities in the Emergent Professions of Artificial Intelligence (AI) and Data Science. *New Technology, Work and Employment* 38: 391–414.

3

WOMEN IN S&T PROFESSIONS

Introduction

This chapter focuses on women in STEM professions other than academia. It first provides an overview of the participation of women and specifically, the educated urban women in the labor force. This forms a backdrop for analyzing gender inequality in some of the key STEM professions, such as engineering, particularly the Information Technology sector, medicine, and STEM entrepreneurship. What is the role of patriarchy and a gendered labor market system in creating gender inequality? How does the STEM workplace interact with these systems? Is the nature of biases different for different STEM professions or are there similarities? These are some of the questions that this chapter attempts to answer.

Female labor force participation, education, and impact of pandemic

The female labor force participation, or FLFP, (as percentage of the total labor force) in India in 2023 was at 28.7%, which is much lower than the world average at about 40%[1] (Table 3.1). The female labor force with advanced education (% of female working-age population with advanced education)[2] in India was higher at 39% in 2023[3] compared to the labor force participation (LFP)-female (% of female population ages 15–64) at 30%, though much lower than men with advanced education[4] (Table 3.1). A comparison with previous peaks suggests there was little improvement in the last two decades.

A combined effect of the pandemic and a gendered labor market took a greater toll on women than on men. Women bore the immediate impact of lockdowns, constituting 73% of job losses in April 2021 (Nikore 2022).

DOI: 10.4324/9781003596660-4

TABLE 3.1 Female labor force participation

	India	*World average (2023)*	*India (previous year peak)*
FLFP (percentage of total labor force)	28.7% (2023)	40.1% (2023)	26.8% (2005)
FLFP (percentage of female population ages 15-64)	30% (2022)	53% (2022)	31% (2000)
FLFP (percentage of female working age population with advanced education)	39% (2023)	NA	36% (2005)
Labor force with advanced education-male (percentage of male working age population with advanced education)	85% (2023)	NA	87% (2000)

Source: World Bank Group data

Among those in the workforce prior to the pandemic, women were seven times more likely to lose work during the nationwide lockdown and highly educated female workers were more vulnerable to job loss than men (Abraham et al. 2022).

The performance on gender parity indices remains lower than in other nations. India is among the lowest at 142 (out of 146 countries) in economic participation and opportunity, and 122 in the proportion of female workers as professional and technical workers at 33.1% (WEF 2024). Overall, India ranks a low 129 in the global gender gap index in 2024, lower by two places since 2023 with the gender parity score not yet reaching the pre-pandemic levels. There is some improvement in the share of women in professional and technical roles, but their share as senior officials, managers, and legislators has dropped. Gender inequality exacts a heavy cost. Bloomberg estimates that closing the employment gap between men and women is likely to expand India's GDP by one-third by the year 2050.[5]

Gender discrimination in labor market

Gender discrimination in the labor market is deep-rooted. Reports suggest an average earning gap of about 4000 INR in regular employment and an overall 27% pay gap between men and women (Oxfam Report 2022: 33).[6] An Oxfam Report (2022) found that women in India, despite the same educational qualifications and work experience as men, will be discriminated against in the labor market due to societal and employers' prejudices. According to a Monster Salary Index survey (2017: 22–23),[7] the pay gap widens as one gains experience and women tend to be underpaid as much as 30% in supervisory positions.

The gender pay-gap is higher in most of the S&T sectors than in others. As per the surveys (Hewitt 2016; MSI 2017), it is the highest in the manufacturing and technology sectors, while it is lowest in the FMCG/FMCD (fast moving consumer goods/durables) sector. The pay gap is higher in the sectors with a lower proportion of women in the workforce (Hewitt 2016).

In general, women in leadership positions continue to be low in proportion as compared to men. Women are increasingly visible at entry and mid-level positions, but the share of women in senior and managerial positions is less than one-fifth (WEF 2024; EY Report 2022; LinkedIn 2024). However, in mid-market companies, the proportion of women in senior management positions is one-third (Grant-Thornton 2024). India has made significant progress in enhancing women's representation on boards, from 6% in 2013 to 18% in 2022, owing primarily to an Indian Corporate Law mandate that imposes there be one woman on the board of all listed companies (EY Report 2022). An EY Report (2022) notes that nearly 95% of the NIFTY 500 companies now have a woman on the board of directors. However, less than 5% of companies have female chairpersons.

Patriarchy and women's work

The gender inequality in the labor market is buttressed by gender-role expectations at home. Women often opt out of employment due to marriage (Fletcher et al. 2017). A large number of women among the upper castes 'choose' not to work due to 'family reasons', there is a lack of safety associated with traveling and timing requirements of jobs, and 'societal norms' and practices that associate respectability with staying out of the workforce for women (Oxfam Report 2022). Employment among women is strongly impacted by the employment status of the mother-in-law, and a woman with an unemployed mother-in-law is more likely to be unemployed herself (SWI 2023). The unemployment among women affects their status in society as there is a strong relationship between women's empowerment and employment (Jain 2018). Patriarchy and a gendered market have affected women's employment after the Covid pandemic. Married women were less likely to return to work and married men more likely to return to work after the nationwide lockdown due to the pandemic (Abraham et al. 2022).

Since late 2021, there has been a 10–15% rise in women job seekers, particularly due to a rise in women's proportion in STEM education; however, a majority of them are seeking work from home (WFH) positions (Business Manager 2022). WFH, which picked up after the pandemic, is enabling greater gender diversity and inclusion as it helps to take care of family constraints (Bhattacharya 2022). Women professionals working from home during the lockdown reported higher work productivity than those working from their workspaces, perhaps owing to the greater satisfaction derived from relational experiences than individualistic goals in the Indian society; although, having children was a constraint on productivity (Kundra et al. 2023).

Gender at the workplace

Literature on professional women's experiences in India indicates a gender-biased environment at the workplace. Women believe that they have less chance than men of receiving working facilities, promotions, and salary increases (Kundu 2003; Kundra et al. 2023), and that organizations discriminate regarding job advancement, annual compensation, and job assignments (Sharma and Kaur 2019). Even if performance management and reward structures are perceived as gender-free, women managers still report that extra hard work is required from them to prove their worth (Nath 2000). Basu (2008) highlights the prevalence of gender stereotypes in Indian corporates and that lack of interaction due to gender segregation accounts for the persistence of most of the stereotypes regarding women. Men employees perceive women professionals to be indecisive, weak, risk-averse, and unwilling to relocate and make sacrifices required of top managerial roles (Jain and Mukherjee 2010). Women employees are rated by men as being less qualified, less competent, and less productive than women rated themselves (Kundu 2003). Although the underlying stereotypes are well-embedded at the workplace, Indian men are in denial about the existence of a glass ceiling (Jain and Mukherjee 2010). Further, professional women lack employer and co-worker support for the use and implementation of family-friendly policies (Vyas 2023).

This chapter analyzes the position of women in some of the key professions related to STEM such as engineering, particularly women in the Information Technology sector, in medicine, and STEM entrepreneurship.

Women engineers in the workforce

A male dominance in technology organizations and a patriarchal cultural context construct unequal power relations and hegemony of men. In most countries, since engineering is male dominated and gender-typed, the beliefs about male competence are buttressed by the presumptions of their specific gender skills (Ridgeway and Correll 2004), thus, raising questions about female competence. In India, competency issues arise not only out of gendered beliefs about skills, but are also closely related to the workplace conditions that disadvantage women. As a result, some engineering sectors are more 'masculine' than others.

The proportion of women engineers is higher as degree holders than as professionals. Women's share of undergraduate engineering degrees is about 29%, and of master's about 35.5% (AISHE Report 2021–2022: Table 35, 36). The proportion of women in engineering professions is about 20% (WEF 2020: 40). More women are employed in computer-related jobs than in the core branches of engineering. Overall, while there are 22% women engineers in IT service companies, women constitute only about 7.7% to 10% of the workforce in many core engineering industries (NER 2016); about 3% in core jobs (12% in tech) in the manufacturing and tech-based heavy sector (SWE 2024); and about 1.4% as engineers/architects in the construction industry (PP-WTC report, 2023). Further, more women than men engineers are estimated to be unemployed, and an older survey (the only available) in the 1990s, estimated it to be five times that of men (Parikh and Sukhatme 2004).

The gender discrimination against women engineers at the workplace is evident in the gender stereotypes and gendered assumptions. Employability of male and female engineers is almost equal (NER 2019). However, in a study of women engineers at the workplace (Rincon et al. 2019), women believed that they had to provide greater evidence of their competence to prove themselves; that they had to walk a tightrope between being too aggressive or too submissive; that due to assumptions of domestic role, they faced negative perceptions about commitment, and exclusion from informal networks; and that junior women felt that senior women adopt 'masculine' aspects of leadership.

Certain engineering careers, such as a hydropower engineer, are more masculine than others. A career as hydropower engineer requires risk-taking, a quality believed to be inherently present in men who are regarded as more capable of facing tough working and living conditions; since organizational success depends upon coping with such conditions without complaining, the gender hierarchy is reinforced, rendering the question of being gender-sensitive as irrelevant (Shrestha et al. 2019).

The Indian official surveys of higher education (or AISHE Reports) list 'architecture' as part of engineering, and it is a female-dominated branch of education with women constituting 58% of the degree holders in architecture (AISHE Report 2021-2022, table 35). As professionals, women constitute a lesser proportion at about 47.6% as architects registered with the Council of Architecture (COA).[8] But this proportion is quite striking compared to other countries, for instance, women constituted only 29% of the registered architects in the UK 2019;[9] about 40% in the US.[10] However, the proportion of women in India practicing architecture in 2023 was only about 20%, less than half of those who are registered.[11]

Though large numbers become architects, Woods (2017) found that women architects in India face discrimination and micro-aggressions at the workplace similar to their counterparts elsewhere and they seem to leave after a few years in the profession. In fact, the profession is seen as a 'male world'.[12] More research is needed to understand gender and architecture. It could be hypothesized that entry into the field might be due to the significance of professional degrees among the urban, middle-class families; while the launching in one's career hits the roadblock of a masculine culture. Perhaps the association of architecture with drawing that is stereotypically seen as a feminine subject by families (Mittal 2021), aids them at the entry stage.

Engineers from various branches (such as ECE, EE, CE, ME) are eligible to be employed in the technical segments of the Information Technology (IT) industry because of their basic knowledge of software (Gupta 2015). Thus, a study of gender and engineering is closely tied to the study of gender and IT.

Women in IT: the growth and challenges

Participation of women

Women's participation in the IT sector seems to have become *stagnant*, although higher than in most sectors and many other countries. Women constituted 21% of the total IT workforce in 2001, increasing to about 30% in 2012 as per the NASSCOM data (Gupta 2015), and still, at present, 30% in 2023 according to a CFA report.[13] Women's participation in the IT workforce is the second highest among all non-agricultural sectors in the country after e-commerce and retail (Women and IT Scorecard 2018[14]).

This participation in IT is also higher in comparison to the participation in the sector in other countries. For instance, in 2022, women held 24% of all computing occupations in the US although they make up more than half of the US workforce, and the percentage of computing occupations held by women has been declining in the US since 1990, when it reached a high of 34%.[15]

The increase in numbers of women enrolled for engineering education is due to a considerable growth in demand for engineers in the IT industry since the 1990s liberalization of the economy. The IT sector is the biggest employment generator and has spawned the mushrooming of several ancillary industries such as transportation, real estate, catering, security, and housekeeping, etc. Direct employment in the IT services and the BPO/ITeS segment has increased from 3.97 million in 2017–2018 to 4.47 million in 2020–2021, according to government estimates.[16] It has been a consistent driver of the economy, accounting for 7.4% of the GDP in 2021–2022[17] and has recorded a 15.5% growth (highest ever), generating $227 billion in revenue in 2021–2022.[18]

The IT advantage for women

The growth of women in the IT sector has been aided by the fact that computers enjoy a woman-friendly image and fit into the overall gendered socio-cultural context. As mentioned earlier, specialization in computers is regarded as suitable for women (Parikh and Sukhatme 2004), the jobs associated with computers are considered 'safe' for women (Varma 2010) and offer a career in software which commands the highest social approval (Bagchi 1999). The changing nature of employment from 'work from office' to 'hybrid' work or 'work-from-home' after the pandemic works in favor of women in computer science and engineering, as their profession has a greater chance of WFH than in some other branches of engineering, such as in civil or mechanical. A woman being skilled in computers is not regarded as abnormal, which is also reflected through Indian cinema (Pal 2010: 7).

The socio-cultural context of a boost in women's participation is supported by the organizational context of the IT professions. The IT sector offers white-collar jobs with a comparatively high salary, easy international mobility, gender-neutral policy based on knowledge-centric skills, flexible work routine, and physically less demanding work processes in a comfortable indoor work environment (Kumar 2001; Upadhye 2006; Shanker 2008). Some other factors that have led to this trend include transportation facility, parental leave, anti-harassment measures, healthcare, and an emphasis on recognizing and supporting women's needs (especially mothers') (Women and IT scorecard 2017).

The IT industry, a relatively new industry (about 30 years old as of this writing), is unencumbered by a traditionally masculine culture. It has been the first to recognize the significance of diversity and inclusion of which gender has been a major component. Gender diversity is seen as a business imperative, particularly since IT firms operate in a highly competitive environment and talent acquisition and its retention are important; also, diversity is viewed as a key to innovation (Buddhapriya 2013). Wipro Technologies launched its Women of Wipro (WoW) initiative, as part of the Wipro Diversity Council, in 2008 to reduce the attrition of women employees and facilitate increased numbers of women managers at a senior level. Infosys has several programs for gender diversity, including those targeting people with career breaks, providing lactation rooms, interaction areas, and workstations to female employees (WE 2023). It won the first NASSCOM-India Today award for gender inclusivity in 2007 and an American Society for Training and Development award in 2009 (Buddhapriya 2013).

The gender diversity initiatives of the firms have helped to enhance the representation of women in the workforce. For instance, the number of women working at the top five IT organizations (Tata Consultancy Services, Infosys, Wipro, HCLTech, and LTIMindtree) jumped 44% between 2020 and 2023 (accounting for 35.4% of the total in 2023) and by 6.5% in just a single year, 2022–23, outperforming the rise in the total overall headcount in these companies (WE 2023; Mani 2023). In 2012, only about one-fifth of the companies had more than 20% women at senior level; by 2015, the number of such firms had increased to one-third (Women and IT Scorecard 2017). The number of companies achieving 30% at the C-Suite level has been increasing by 2% a year since 2012 (Women and IT Scorecard 2018). Diversity has shown good business results as well. Corporations with at least 10% women on company boards have 2.5–5% higher returns on equity, and firms where women are at least 30% of C-Suite have 15% higher profitability than others (Women and IT Scorecard 2017).

Nevertheless, there are several indicators of gender inequality in the sector.

Gender construction in the IT sector

The stagnation in women's participation in the IT sector at 30% since 2012 suggests gender inequalities are hampering the growth in this field. Skillsoft Report (2022), based on a survey of IT female professionals across the country, found that there is a lack of gender equality in opportunities, that the work of men gets highlighted more than that of women (70% respondents) and 43% of women respondents faced harassment at work. An overall underrepresentation of women in the sector was also felt to be a challenge, particularly in larger companies. Many reported inequity, condescending treatment, and skepticism in their abilities.

Studies have found that gendered assumptions are built into competencies required for work in the IT sector. For instance, women software professionals are provided with less critical and complex projects and often find themselves as the lone woman on a team, with immense pressure to perform, while more trust is given to the male professionals (Maji and Dixit 2020). Assumptions, such as that men have greater 'working flexibility' and potential for 'working under pressure' are built into the appraisal system, so that women find evaluation biased and stressful (Arun et al. 2007). The employers perceive that women employees might not be able to combine work with family/childcare commitments (Women and IT Scorecard 2017). Leaders deny equal opportunities to women as the definition of 'merit' continues to hold masculine attributes (Chakraborty 2019). Within teams there is a gendered division of tasks, where women participate in communications and relational aspects and men perform the logic and technology, which affects career growth and self-motivation (Maji and Dixit 2020). Thus, gender inequalities persist in the seemingly gender-neutral industry.

Further, women appear to be lagging behind in the emerging technologies. Technology is making rapid strides as new areas have emerged such as

Artificial Intelligence (AI). However, according to the data collected by the social networking site LinkedIn and published in the Global Gender Gap Report, India has only 22% of professionals working in AI, similar to the world average (UNESCO Science Report 2021: 7). Cybersecurity is another emerging area, ranked highest in salary and as the top investment area by IT decision-makers but, only 1% of women hold positions in cybersecurity (Skillsoft Report 2022).

The IT industry is experiencing a considerable gender pay gap. Skillsoft Report (2022) indicated that the lack of equal pay was a major area of concern for female software professionals. The gap exists at all levels, lowest for three to five years of experience level and highest for six to ten years of experience, as reported by Monster Salary Index (MSI-IT 2016: 26–17).

The IT industry exhibits vertical segregation. A recent report (CFA 2023) shows women as only 8.3% of the key management personnel in the IT sector. Another report found that while about 51% of entry level recruits are women, over 25% of women are in managerial positions, and less than 1% are in the top level/C-Suite (Women and IT Scorecard 2017). Thus, there is stagnation at the entry- and middle-career levels, and the gender gap widens at the senior levels significantly. There is gender gap even in the new technologies within the IT sector. For instance, the gender gap in GenAI ranges from 35% in junior positions to as high as 63% in senior management positions (Jain 2024). This indicates that a growth in numbers of qualified women engineers does not imply a consecutive decline in gender segregation in the labor market.

There are further evidences of gender inequalities in the profession. Maji and Dixit (2020) found that there are few women visible above the age of 30 years in the IT Industry, which suggests that a large proportion of women tend to exit from the industry after the first few years of employment. While men and women start their careers in IT companies in India at similar ages, men at senior positions are often younger than women at a similar level. This indicates that women with career breaks or part-time work take longer to move up (Women and IT Scorecard 2017).

IT companies face a significant problem of retention following maternity leave (Women and IT Scorecard 2018). The two most common constraints perceived by IT companies in employing women are: a gendered assumption that a woman's primary responsibility is towards her family, thus, wondering whether women can combine work with family commitments, or will she leave job once married; and, secondly, the challenge of dealing with government's regulations regarding working hours and parental leave (Women and IT Scorecard 2017; Patel and Parmentier 2005; Maji and Dixit 2020).

These findings reflect on the general status of women in the STEM workforce in the private sector. First, the work-family balance issue is significant for the recruiters; second, government regulations as to a work-life balance are perceived as a burden by companies; third, high attrition among women following maternity leave bothers the companies but remains an unresolved issue for the most.

The attrition partly appears to be a product of a gap between the policymakers and those who implement the policies. Although flexibility and work-life balance programs exist in organizations, they are not readily available to women in practice (PwC 2016); use of policies is hampered by a lack of organizational support (Vyas 2023); and taking advantage of flexibility and work-life balance programs is believed, by women, to have negative consequences for their careers (Women and IT Scorecard 2018). There is considerable evidence of such an implementation gap between policy and practice elsewhere also (Holt and Lewis 2011; Gambles et al. 2006). Policies are often undermined by non-supportive managers (Lewis, Brannen and Nilsen 2009) and gendered workplace cultures (Haas and Hwang 2007); there is a gap between the discourse of supporting women and the reality of cultural and structural barriers (Lewis and Humbert 2010).

Further, Chakraborty (2019) found that the discourse on business case for gender diversity uses gender stereotypes. Often, initiatives for inclusivity demonstrate or digress into gendered measures. Replacement of a male incumbent as the Human Resources (HR) Head by a female might be a part of D&I strategy of the companies (in various sectors) since women are believed to have feminine qualities of empathy, emotional intelligence, and interpersonal skills (Singh 2018).

After the pandemic, employees prefer WFH or a hybrid form of employment, while IT companies prefer workers to WFO (work from office) to avoid moonlighting.[19] There are indications that women prefer WFH or flexible work as it helps to overcome family constraints.[20] Thus, there is a gap between the perspectives of individuals, particularly of women and companies.

The IT workplace reproduces socio-cultural stereotypes in the informal work environment. For instance, success of women might be attributed to their sex status, the usual stereotypes about women form the content of jokes (Maji and Dixit 2020) suggesting microaggressions. The gendered nature of socialization that discourages male-female interaction carries forward at the workplace. Having frequent conversations with a male colleague is perceived as over-friendliness, and the woman's career growth might be attributed to such behavior (Maji and Dixit 2020). These authors indicate that a lack of interaction translates to less networking, greater acceptance of injustices, and less aggression that harms career; and, as a reaction to masculine culture, some women technology leaders adopt masculine behaviors.

The glass ceiling in organizations affects leadership opportunities in various ways. The opinions of colleagues (male mostly) that international projects are unsuitable for women due to the latter being family-centric, impacts opportunities and growth (Maji and Dixit 2020). International assignments, which offer an opportunity for skill development and career progression, are reported more for women than men, but they are usually undertaken by younger women (Women and IT Scorecard 2018). The upskilling programs organized by IT companies are not sufficiently focused to promote women to leadership positions; and share qualification (companies having paid–up share capital of one thousad million rupees or more) for female directorship prescribed by listed public companies is an impediment for women to be considered for executive positions (Pallissery et al. 2023).

The nature of an organization also impacts women. A small company may provide its women employees a better work environment than a big organization (Gupta et al. 2015). This is also indicated in a NASSCOM Report (Women in IT leadership in India 2018) according to which the size of a company matters and the smallest companies are most gender diverse.

However, employment in the IT sector tends to empower women and challenge patriarchy. Kelkar et al. (2002), in a study of Bangalore and Delhi, found that the nature of work in the IT industry, with its networking capacities, tools (such as e-mail and the internet), and individuation of these capacities, has resulted in greater women's capacity to make decisions and thereby enhanced their agency. Further, prestige attached to the IT industry allows young women to work night shifts, live alone in cities, and control their incomes. The high salaries of women software engineers lead to an accumulation of 'symbolic capital' in terms of greater respect from their husbands than they would otherwise have, making parents proud to have a working daughter who can often support them just as well as a working son can (Radhakrishnan 2008). The wife supporting her parents challenges the patriarchal norm of a woman as peripheral to her parental family upon her marriage as she shifts residence and allegiance to her husband's family.

Gender segregation in the labor market does not seem to deter young women engineers from pursuing a degree, and subsequently a job in the IT industry. However, the growth of women in computer science UG enrollments is much slower than that of men's (as noticed in chapter 1). Their proportion in IT jobs has not risen since the last decade. Women seem to be losing out in the competition for education as well as jobs in this field.

In order to halt the trend of declining or stagnant growth in women's proportions in education and jobs in IT, urgent measures need to be thought of. This is because the supply of engineers and acquiring talent is critical for business and innovation, which requires tapping the talent of women professionals. Opportunity structure for women, including upskilling in the current scenario of emerging technologies, needs to be scaled up. Organizations need to focus on the real challenges for women, such as the need for childcare centers, elderly care/daycare structures, mobility, and greater integration of women's concerns with the opportunity structure.

Medicine: 'woman-friendly'?

Medicine is a profession to which women have had access historically, and in which they are presently a near parity in the proportion of degrees earned. However, as practicing professionals, they are a minority. While there are about 0.80 million allopathic doctors registered in India, the proportion of females among doctors is only about 29%, and among dentists only 37% (Karan et al. 2021: Fig. 4, estimated from NSSO 2017-18). According to a WHO statistic, in 2020, women were only 14.2% of the allopathic doctors.[21] These figures as to the representation of women doctors are surprisingly low, considering that the proportion of female allopathic degree holders (MBBS) is about 49%, and that of dentists (BDS) is 73% (AISHE 2021-2022; Table 34); female degree holders of Master in Medicine (MD) are about 53%, and in dental surgery (MDS) 68.5% (Table 34a). Overall, the density of physicians in India (0.7) in 2020 was much less than the world average (1.7).[22] Women with degrees in medicine constitute a huge talent pool that remains untapped.

Clearly, a large proportion of female doctors are inactive or probably have meandered into different career paths or migrated abroad. About 31% of women with a technical education in medicine are out of the labor force and among medically trained 30- to 40-year-old women, 20% are not working and most of them are engaged in household work instead (Gulati et al. 2023), implying a considerable loss to society. Women fail to have careers in medicine due to various bottlenecks in the system.

The bottlenecks to a medical career are related to both the issues of a work-life balance and workplace discrimination. The long work hours and juggling work-home responsibilities could be daunting in a medical career. Women physicians in the public sector are granted maternity leave/childcare leave as per GOI rules. Reliable daycare is available in few hospitals/medical colleges for toddlers but none for school-going children. There is a lack of opportunity for flexible training or for creative scheduling of a job or training if women want to come back to an academic or professional life after a gap in one's career (Sood and Chadda 2010).

Medicine is male dominated through hegemony over positions of power and structural dominance. Thus, the situation is not very different in India compared to many other countries (Ramakrishnan et al. 2014). In the statistics available in NIRF 2018 (but not in subsequent NIRF reports), women constituted 26.7% of the faculty at the reputed AIIMS, and about 31% in PG Institute of Medical Education and Research at Chandigarh in 2017–2018.

There are few women physicians in senior positions. As of the date of this writing, AIIMS has had only one female director, and ICMR only two women as director general.[23] Sood and Chadda (2010) found that 36% of the women faculty in AIIMS were in senior faculty positions compared to 64%

in junior faculty positions and only 8% of the women faculty were employed as departmental chairs.

Surgery is male dominated and is construed as a masculine profession (Bhadra 2011). The proportion of female surgeons is only about 2.8%,[24] although their proportions as Master in Surgery (MS) degree holders is about 46.5% (AISHE Report 2021–2022, table 34a). There are fewer women cardio, neuro, cancer, and transplant surgeons compared to those in gynecology, pediatrics, cosmetology, dentistry, and dermatology, which are seen as women-friendly (Ghosh 2023). Surgery is believed to be physically demanding and combining professional career with family responsibilities is considered difficult (Sood and Chadda 2010). Women also encounter a trust deficit and a resistance from male colleagues, thus, requiring considerable hard work so as to achieve recognition for their skills (Ghosh 2023). Teams are generally male dominated and usually led by men. A survey of Indian oncologists (Bajpai et al. 2020) found that 58% of respondents worked in male-majority teams, and only 15.4% of women participants worked in female-majority teams and about 67% of respondents (male and female) had male managers.

In medical academic centers, invited talks and participation in national and international committees and conferences are important for promotions. However, both require travel, which challenges women with family commitments, and such venues are gender-blind having no support for such women (Bajpai et al. 2020). For women medical faculty, combining academic pursuits, clinical responsibilities, and domestic roles is a difficult task, requiring greater sensitivity by the system. Many issues faced by women faculty are similar to those faced by women faculty in other STEM fields, such as male bias in hiring/promotion, considering women as liabilities due to their family commitments and regarding men as natural leaders (Bajpai et al. 2020).

As researchers, women constitute 26% of the total scientific manpower in the Indian Council of Medical Research (ICMR), a premier medical research organization (Desai-Gorakshakar et al. 2011). Women researchers in medicine are usually the lead authors in only 26% of the top Indian journals as found in the case of oncology (Bajpai et al. 2020) and neuroscience (Dubey et al. 2016). Findings elsewhere regarding medical journals (e.g., Broderick and Casadevall 2019) have shown female authors not receiving proper credit. The male dominance of medical publications has certain narrow exceptions. Thelwall et al. (2019) found that in India the proportion of female authors per male author varied from 0.05 (orthopedics and sports medicine) to 1.23 (obstetrics and gynecology).

The measures suggested by Bajpai et al (2020) for supporting women oncologists could well apply to most female faculty in medicine. These include flexible educational courses, mentorship programs, online professional career development tools, network of women in the same specialization, family-friendly facilities at conferences, dedicated seats for women in national and international committees, and soft-skills training.

There are stereotypes distinctive to medicine. Kumari et al. (2015) indicate that a majority of the women doctors feel that patients discriminate between male and female doctors with men being preferred for consultation, suggesting that male doctors are perceived as superior to women doctors. In India, the medical education has been found to be divorced from the gender perspective, and gender biases are actively promoted in medical textbooks used for teaching (John 2016). Gender sensitization is essential in curriculum because it has consequences for diagnosis and treatment.

Most female doctors are concentrated in the urban areas. According to Rao et al. (2011), only 6% of the allopathic doctors in rural areas are women. However, in general, the number of allopathic doctors in urban areas is four times that in rural areas; about 74% of the graduate doctors live in urban areas, serving only 28% of the national population. (Thayyil and Jeeja 2013). Probably there is a need to tap the talented female and male students in the villages, who will be more devoted towards building medical careers in villages.

Thus, similar to architecture, women in medicine obtain education in large numbers, but few practice the profession, suggesting, one, that degrees are more important than successful careers for women's families; and, second, gender discrimination in the labor market and at the workplace. However, unlike architecture, which is viewed as a 'male' domain, medicine is viewed as a 'female-friendly' profession, owing probably to certain branches of medicine, such as gynecology and obstetrics, that are woman-centric.

Women led tech start-ups: new hope?

Although India ranks low in women's entrepreneurship (ranking 57 out of 65 countries in women's entrepreneurship: MIWE 2021), there are encouraging figures for women's tech entrepreneurship. There has been a phenomenal 11 times increase in the number of women entrepreneurs in tech in the last decade (2010–2020) (Makers India report 2020). India stands out for its high proportion of innovative entrepreneurs among women as indicated by a high TEA (Total early-stage Entrepreneurial Activity) or start-up rate among women at 11.5%, compared to the world average of 10.4% in 2022; and by almost parity in the female/male TEA ratio at 0.98 (GEM Report 2023). Women's entrepreneurship in general has seen an increase. While in 2010–2011, the participation of females in the formal entrepreneurship sector in the country was less than 10% (WISAT 2012), in 2022 women's share in start-ups with at least one female founder/co-founder was 18% (Nasscom-Zinnov 2022).

Overall, India has a large presence of technical start-ups and continues to retain its position as the third largest tech-based start-up hub in the world (Nasscom-Zinnov 2022). Tech start-ups are closely linked to innovation. India has improved its rank in the global innovation index (WIPO 2023), from 46 to 40 (out of 132 economies) from 2021 to 2023; and, has risen to sixth position in the filing of patents in 2021 (WIPO-IP 2022). Despite the fragile economic scenario since 2020, the tech start-up ecosystem remained strong with about 23 unicorns emerging in 2022 alone (Nasscom-Zinnov 2022). The government support for women entrepreneurs during the pandemic has been among the strongest in India (GEM report 2022: 89). Tracxn found that in 2023, women entrepreneurs' share of start-ups in the Indian tech industry exceeded 18%; India ranks second after the US in terms of funding raised by companies with women founders in the tech start-up ecosystem (Jaswal 2024). Women's participation in innovation is considered significant for achieving economic targets. The government recognizes that 'high growth rates of 9–10% year after year is not possible for India until women don't become key entrepreneurs'.[25]

However, the gender divide is considerable. Women-founded unicorns are only 15% of the total unicorns with a maximum representation in the e-commerce sector (Inc42 2022). Women-led start-ups receive a miniscule proportion of the overall funding raised by the start-ups; in fact, there is a decline in the share of women-led start-ups in overall funding from 9.5% in 2022 to 5% in 2023 (Inc42 2023). India was among the nations that are at the bottom in the proportion of women inventors listed in patent applications in 2021 at 10.2%, compared to the world average of 16.5% (WIPO-IP 2022).

Women's entrepreneurship faces several normative and systemic constraints. Women in STEM are mostly a middle-class urban category, and this class perceives entrepreneurship as risky (Niti Aayog 2015; Pandey and Pansera 2020). Financing at the initial stage is the most serious impediment for women (Korreck 2019); other issues include stereotypes, work-life balance stress, traveling, odd/long working hours, and risk-averseness (Shukla et al. 2018; Gupta and Etzkowitz 2021). Women usually do not have collateral to offer and families hesitate in financially supporting a daughter's venture. Notoriously unsafe cities and public spaces, a lack of childcare facilities coupled with the social perception of (entrepreneurial) mothers as selfish (Korreck 2019), also inhibit women entrepreneurs.

Constraints for women include the still male-dominated angel investing community. Women find it difficult to network, build business relationships, get customer leads, and access talent (Korreck 2019). Women's participation in tech entrepreneurship and in entrepreneurship in general, is an extension of women's position in the labor force. First, women's status as primary caregivers interferes in obtaining funding. Many female entrepreneurs have reported that during fundraising meetings, they often get asked questions that are personal and usually not related to their respective businesses (Balaji 2017). Further, sustaining and scaling up the business is difficult because the ecosystem is still heavily male dominated (Global Women Entrepreneurs Leader Report 2015: 15). Industry networks are masculine, exhibiting traits such as, 'bro-culture', aggressive competition, and sexual harassment (Korreck 2019). Yet, paradoxically, young women entrepreneurs seem to deny bias (Korreck 2019). Familial support and developmental relationships are critical for urban women entrepreneurs (Chaudhuri, Ghosh, and Abichandani 2020).

Studies worldwide have shown that while academic careers in science and technology encounter resistance to gender equality, start-ups provide new career opportunities for women in STEM due to a gender-neutral environment of accelerators (Etzkowitz 2013). In India, women's participation in research and academic careers is low, with women faculty in Indian scientific institutes constituting about 13.5% of the total;[26] and women in R&D at 18.6%, including the industrial sector (S&T Indicators 2023). With the proportion of women postgraduates and doctorates in STEM at about 40%, women constitute a substantial pool of potential to be tapped for research-based entrepreneurship, offering them an option other than traditional academic careers. While some of the greatest strengths of India lie in entrepreneurship policies and availability of graduates in science and engineering, one of its greatest weaknesses lies in the low proportion of women employed with advanced degrees (WIPO 2022). It is in this context that academic incubators become an important area for study.

Academic incubators

An academic incubator is conceptualized as a hybrid organization in the 'triple helix' of interaction among university, industry, and government to support the start-up process (Etzkowitz et al. 2005). Academia in India has emerged as a partner for entrepreneurs and a major source of implementing government schemes for start-ups through incubators. Academic incubators in an academia provide networking with researchers, alumni, industry, and markets and enable technology transfer, channeling of funding, provide infrastructure, and facilitate research-based product development.

Entrepreneurship involves risk-taking and while the traditional concept of an entrepreneur as an individual risk-taker might be a myth, the institutionalized support structure of an academic incubator offers a de-risking process (Etzkowitz 2003). A study of female high-tech entrepreneurs was conducted at an academic incubator at an IIT (Gupta and Etzkowitz 2021). This study found that often the realization by women that the technology conceptualized by them could be turned into a commercial product occurred while pursuing research in academia. A triple support system of academia, incubator, and the government aided the emergence of an opportunity. Availability of funding through government schemes supported by the academic incubator provided a launch pad for them. The incubator created an internal support structure, providing mentors, networks, financial, and legal help to the participants to develop their venture. Further, the incubator, through formal and informal networking, assisted in countering barriers related to a mixed-sex interaction and homophily. Finding team members and managing them could be challenging for the start-ups. However, they often tapped in to the placement process of an academic institute. Within their own company, these women entrepreneurs consciously adopted egalitarian behaviors and created their own culture.

However, there are issues beyond the domain of incubator and the women entrepreneurs might face the challenge of masculine notions of 'leadership' or traveling alone for networking. Further, there are gendered-role expectations, and in challenging those, women entrepreneurs create agency for themselves (Gupta and Etzkowitz 2021).

Challenging gender stereotypes

The women entrepreneurs challenge the stereotype that only men have the ability required to venture in to high-tech entrepreneurship. As found elsewhere (e.g., World Bank 2018), and in the case of Indian women entrepreneurs in general (Samantroy and Tomar 2018), this study (Gupta and Etzkowitz 2021) also showed that entrepreneurship strengthens women's agency in terms of their ability to challenge social constraints. In this aspect, they are often supported by their spouses or parents. For instance, support in terms of parents delaying the marriage of their daughter, or spousal financial support and help in the household chores, or ignoring family pressures for starting a family. There is an increasing recognition by women that choice of a marital partner is actually a vital career decision for them.

Government's dedicated platform for women entrepreneurs and various schemes provide funds, mentoring, and networks. However, lack of interdisciplinary education, lag in the educational system which fails to inculcate creativity and curiosity, few role models in the field of women's scientific entrepreneurship, and a lack of strong innovation networks for women entrepreneurs are some of the aspects that hamper women's scientific entrepreneurship (Chatterjee and Ramu 2018). Creating optional courses in higher educational institutes (such as the IITs) on entrepreneurial education and financial/legal skills will immensely help women (and men) entrepreneurs in high-tech. One critical factor for all women professionals in the Indian context is obtaining a work-family balance, which requires strengthening of institutional infrastructure for care-giving.

Conclusions

In general, women professionals confront gendered perceptions regarding their ability, gender pay-gap, difficulties in informal interaction, discrimination in evaluation, and the glass-ceiling. The stereotypes at a workplace and assumptions of a domestic burden on women deny them opportunities. Although the pandemic hit women more than men, they seem to have recovered in terms of economic participation. However, women are willing to shift jobs if WFH and a hybrid mode of working are available (Bhattacharya 2022), which appears to be a course driven by both gendered societal expectations and the workplace not being particularly inviting.

While there are similarities in construction of gender between S&T professions and other professions, there are differences as well. For women engineers, the masculinity of the engineering domain and the male dominance create gender salience at the workplace. They need to provide greater evidence of their competence than men. A gendered division of tasks, networks, and informal environment affects teamwork and career progression. Masculine attributes buttress the male dominance in engineering fields, such as the attribute of 'risk-taking' or 'working under pressure'.

In the IT industry, due to a perceived 'woman-friendly' nature of computer-related branches, a large number of women enter education and the workforce. Although due to the newness of an industry unencumbered by traditional male-dominated cultures, and as the earliest in introduction of gender diversity measures, IT workplaces might be more 'woman-friendly' than many other sectors, yet the promise of gender-equality is far from being realized. The proportion of women in the IT industry has been stagnant since the 2010s at 30%. Women's proportion in the new areas of technology (e.g., Artificial Intelligence, cybersecurity) is much lower than their proportion in this industry. There are few women after the age of 30 and most seem to quit careers within a few years of work. There is a gendered environment in the IT organizations, such as assumptions of women's family responsibilities, denying opportunities for international projects, doubting their ability to handle complex projects, gaps in policy and implementation, and there are hardly any daycare/elderly care centers at the workplace.

In medicine, although there is a near parity in education, their proportion is quite low as doctors. As many as one-third of female MBBS degree holders do not practice medicine. From the few available studies, there appear to be constraints on work-life balance and workplace discrimination. Its continued perception of being 'female-friendly' is perhaps due to a considerable horizontal segregation where some branches are female dominated.

Although STEM professions are gendered, the mere presence of female professionals at the STEM workplaces challenges the masculine space. Further, STEM professions empower women to question patriarchy and modify gendered norms in their private lives. For instance, participation in prestigious STEM occupations enables women to contribute financially to the parental family and to raise their social status or to delay marriage and defy gendered role expectations.

Women's tech entrepreneurship is an emerging possibility for women in STEM. It has increased eleven times between 2010 and 2020. While there are several constraints, an academic incubator, with the aid of government programs, provides a solution to many issues including initial funding, product development, acquisition of customers, and business networks. However, women's tech entrepreneurship is a nascent area requiring greater research.

A prominent STEM profession is being an 'academic'. Academia offers highly prestigious research careers in STEM. The next two chapters analyze challenges to gender equality in science academia.

Notes

1 World Bank Data: Labor force, female (% of total labor force) – India. Available at: https://data.worldbank.org/indicator/SL.TLF.TOTL.FE.ZS?locations=IN; Labor force participation rate, female (% of female population ages 15–64)-India. Available at: https://data.worldbank.org/indicator/SL.TLF.ACTI.FE.ZS?locations=IN-1W
2 Advanced education comprises short-cycle tertiary education, a bachelor's degree or equivalent education level, a master's degree or equivalent education level, or doctoral degree or equivalent education level according to the International Standard Classification of Education 2011.
3 World Bank Data: Labor force with advanced education, female (% of female working-age population with advanced education) – India. Available at: https://data.worldbank.org/indicator/SL.TLF.ADVN.FE.ZS?locations=IN
4 World Bank 2023: Labor force with advanced education, male (% of male working-age population with advanced education) – India. Available at: https://data.worldbank.org/indicator/SL.TLF.ADVN.MA.ZS?locations=IN
5 Bloomberg 2022: https://www.ndtv.com/india-news/covid-cut-indias-women-out-of-job-market-90-not-in-workforce-report-covid-pandemic-affects-job-market-women-lose-jobs-in-covid-3031243).
6 Oxfam India's 'India Discrimination Report 2022' is based on Government of India data on employment and labor from 2004–2005 to 2019–2020.
7 Monster Salary Index survey: Monster Salary Index 2017: https://media.monsterindia.com/logos/research_report/MSI_Full_Report_2016_July_2017.pdf
8 Council of Architecture (COA) website, available at: https://www.coa.gov.in/index1.php?lang=1&level=1&sublinkid=36&lid=49
9 In UK: https://arb.org.uk/arb-publishes-its-latest-annual-report-nr20/#:~:text=While%20overall%2071%25%20of%20those,30%20was%20exactly%2050%2F50
10 In US: 'Where Are the Women? Measuring Progress on Gender in Architecture'. Available at: https://www.acsa-arch.org/resource/where-are-the-women-measuring-progress-on-gender-in-architecture-2/
11 Architecture Profession: A Man's World. January 13, 2023. Available at: https://architecture.live/architecture-profession-a-mans-world/
12 Abha Lambah, an architect, 2021, quoted in the article 'Only 20% of licensed women architects in India become practitioners' Available at: https://www.constructionworld.in/policy-updates-and-economic-news/only-20--of-licensed-women-architects-in-india-become-practitioners/29155

13 See Gupta (2015); Also, NASSCOM 2012: Indian IT BPM sector in India 2013 – A Summary | PDF (slideshare.net); CFA Report (2023): CFA analysis is based on a sample of voluntary BRSR (Business Responsibility and Sustainability Report framework) disclosures for FY2022 to provide.
14 Women and IT Scorecard 2018: Survey of 150 Indian IT sector companies and a survey undertaken with 270 Indian IT workers. Women and IT Scorecard 2017: Survey of 55 IT sector companies across Delhi, Bangalore, and Hyderabad.
15 Women's bureau, an official website of the US: https://www.dol.gov/agencies/wb/data/Facts-over-Time
16 Employment generation in ITeS/BPO: https://static.pib.gov.in/WriteReadData/specificdocs/documents/2022/feb/doc202222418201.pdf
17 India Brand Equity Foundation, Department of Commerce, GOI: https://www.ibef.org/industry/information-technology-india#:~:text=The%20IT%20industry%20accounted%20for,to%20India's%20GDP%20by%202025
18 Revenue generation: https://www.meity.gov.in/revenue-trend; https://nasscom.in/knowledge-center/publications/technology-sector-india-2022-strategic-review
19 'Companies want their employees back to office, but it is turning out to be difficult', Business Insider, March 20, 2023. See: https://www.businessinsider.in/thelife/news/companies-want-their-employees-back-to-office-but-it-is-turning-out-to-be-difficult/articleshow/98602147.cms
20 'Women are moving jobs for the work-from-home perk.' Economic Times. March 8, 2022. See https://economictimes.indiatimes.com/news/company/corporate-trends/women-are-moving-jobs-for-the-work-from-home-perk/articleshow/90061209.cms
21 National Health Workforce Accounts Data Portal. WHO. https://apps.who.int/nhwaportal/?AspxAutoDetectCookieSupport=1
22 Physicians (per 1,000 people) – India: https://data.worldbank.org/indicator/SH.MED.PHYS.ZS?locations=IN
23 AIIMS: https://aiims.edu/index.php?option=com_content&view=article&id=629&Itemid=636&lang=en; ICMR: https://main.icmr.nic.in/former-dg
24 Top women surgeons 2021: https://www.womenentrepreneurindia.com/ranking/top-women-surgeons-2021-rid-49.html
25 Amitabh Kant, CEO, Niti Aayog. Speaking at the 'Britannia Marie Gold My Startup' initiative in collaboration with National Skill Development Corporation (NSDC). Available at: https://economictimes.indiatimes.com/small-biz/startups/newsbuzz/women-entrepreneurs-key-to-india-achieving-9-10-growth-amitabh-kant/articleshow/74504999.cms)
26 Biaswatchindia: https://biaswatchindia.com/base-rates-of-indian-women-faculty/

References

Abraham, R., Basole, A., and Kesar, S. (2022). Down and Out? The Gendered Impact of the COVID-19 Pandemic on India's Labormarket. *Economia Politica* 39(April 2022): 101–128.

Arun, S., Heeks, R., and Morgan, M. (2007). ICT Initiatives, Women and Work: Reproducing or Changing Gender Inequalities? In Ghadially R. (ed.), *Urban Women in Contemporary India: A Reader*, SAGE: New Delhi, 297–308.

Bhadra, M. (2011). Indian Women in Medicine: An Enquiry Since 1880. *Indian Anthropologist* 41 (1): 17–43.

Bagchi, S. (1999). India's Software Industry: The People Dimension. *IEEE Software* 16(3): 62–65.

Bajpai, J., Mailankody, S., Nair, R., Surappa, S. T., Gupta, S., Prabhash, K., et al. (2020). Gender Climate in Indian Oncology: National Survey Report. *ESMO Open,* 2020 April, 5(2): e000671. https://doi.org/10.1136/esmoopen-2020-000671

Balaji, S. (2017). Women-Led Startups Could Grow India's Economy By 60%, But Obstacles Remain for Female Founders. *Forbes*, December 14, 2017. Available at: https://www.forbes.com/sites/sindhujabalaji/2017/12/14/india-needs-more-women-led-startups-it-makes-economic-sense/#5334ef9c7e58. Accessed on September 4, 2024.

Basu, S. (2008). *Gender Stereotypes in Corporate India: A Glimpse.* Response Books: New Delhi.

Bhattacharya, S. (2022). *Women are Moving Jobs for the Work-from-Home Perk, The Economic Times*, March 8, 2022. Available at: https://economictimes.indiatimes.com/news/company/corporate-trends/women-are-moving-jobs-for-the-work-from-home-perk/articleshow/90061209.cms?from=mdr. Accessed on September 4, 2024.

Broderick, N.A., and A. Casadevall. (2019). Gender Inequalities among Authors who Contributed Equally. *elife* 8: e36399. https://doi.org/10.7554/eLife.36399

Buddhapriya, S. (2013). Diversity Management Practices in Select Firms in India: A Critical Analysis. *Indian Journal of Industrial Relations* 48(4): 597–610.

Business Manager (2022). *15% Rise in Women Job Seekers Since 2021.* March 9, 2022. Available at: https://www.businessmanager.in/15-rise-in-women-job-seekers-since-2021-finds-spectrum-talent-managements-report/. Accessed on September 4, 2024.

CFA. (2023). *Mind the Gender Gap Analysis of Women's Participation, Pay, and Other Measures in Indian Public Companies.* By Sivananth Ramachandran, Jolly Balva, and Meera Siva. CFA Institute. Available at: https://www.cfainstitute.org/-/media/documents/article/industry-research/Mind-Gender-Gap-Indian-Public-Companies.pdf. Accessed on September 4, 2024.

Chakraborty S. (2019). The Business Case for Gender Diversity in the Indian Information Technology Industry. In S. Nachmias and V. Caven (eds), *Inequality and Organizational Practice.* Palgrave Explorations in Workplace Stigma. Palgrave Macmillan: Cham, Switzerland, 211–233.

Chatterjee, C., and Ramu, S. (2018). Gender and Its Rising Role in Modern Indian Innovation and Entrepreneurship. *IIMB Management Review* 30(1): 62–72.

Chaudhuri, S., Ghosh, R., and Abichandani, Y. (2020). Exploring the Risky Terrain of Entrepreneurship with Support from Developmental Relationships: Narratives from Indian Women Entrepreneurs. *Advances in Developing Human Resources* 22(2): 137–149.

Desai-Gorakshakar, Kala A., Padwal, Vijay G., and Ghosh, K. (2011). Women Scientists in India are Marching ahead. *Current Science* 101(3): 259–261.

Dubey, D., Sawhney, A., Atluru, A., Amritphale, A., Dubey, A., and Trivedi, J. (2016). Trends in Authorship Based on Gender and Nationality in Published Neuroscience Literature. *Neurology India* 64(1): 97–100.

Etzkowitz, H. (2003). Innovation in Innovation: The Triple Helix of University-Industry-Government Relations. *Social Science Information* 42(3): 293–337.

Etzkowitz, H., Carvalho de Mello, J.M., and Almeida, M. (2005). Towards "Meta-Innovation" in Brazil: The Evolution of the Incubator and the Emergence of a Triple Helix. *Research Policy* 34(4): 411–424.

Etzkowitz, Henry. (2013). StartX and the 'Paradox of Success': Filling the Gap in Stanford's Entrepreneurial Culture. *Social Science Information* 52(4): 605–627.

EY Report (2022). *Diversity in the Boardrooms. Progress and the Way Forward.* Ernst & Young LLP: India. Available at: https://assets.ey.com/content/dam/ey-sites/ey-com/en_in/topics/women-fast-forward/2022/09/ey-dei-report.pdf. Accessed on September 4, 2024.

Fletcher, Erin K., Pande, R., and Moore, C.T. (2017). *Women and Work in India: Descriptive Evidence and a Review of Potential Policies.* HKS Faculty Research Working Paper Series RWP18-004, December 2017. Available at https://www.hks.harvard.edu/publications/women-and-work-india-descriptive-evidence-and-review-potential-policies. Accessed on September 4, 2024.

Gambles, R., Lewis, S., and Rapoport, R. (2006). *The Myth of Work-Life Balance. The Challenge of Our Time for Men, Women and Societies.* Wiley: Chichester.

GEM Report (2023). *GEM 2022/2023 Women's Entrepreneurship: Challenging Bias and Stereotypes.* Global Entrepreneurship Monitor: London. Available at: https://www.gemconsortium.org/reports/womens-entrepreneurship. Accessed on September 4, 2024.

GEM Report (2022). *Global Entrepreneurship Monitor. 2021/2022 Global Report. Opportunity Amid Disruption.* Available at: https://www.gemconsortium.org/file/open?fileId=50900. Accessed on September 4, 2024.

Ghosh, R. (2023). Ahead of International Women's Day, A Look at Why There are so Few Women Surgeons in India, *The Indian Express*, March 5, 2023. Available at: https://indianexpress.com/article/lifestyle/life-style/international-womens-day-why-there-are-so-few-women-surgeons-in-india-8478694/. Accessed on September 4, 2024.

Global Women Entrepreneurs Leader Report (2015). 2015 Global Women Entrepreneur Leaders Scorecard. Executive Summary. Available at: https://i.dell.com/sites/doccontent/corporate/secure/en/Documents/2015-GWEL-Scorecard-Executive-Summary.pdf. Accessed on September 4, 2024.

Grant-Thornton. (2024). *Women in Business 2024.* Grant Thornton. Available at: https://www.grantthornton.global/globalassets/1.-member-firms/global/insights/women-in-business/2024/grant-thornton-women-in-business-report-2024.pdf. Accessed on September 4, 2024.

Gulati, K., Davies, J., and González de la Fuente, Á. (2023). Addressing Leadership Competency Gaps and Gender Disparities in India's Medical Workforce: A Call to Action. *Lancet Reg Health Southeast Asia,* July 5, 2023, 16: 100247. https://doi.org/10.1016/j.lansea.2023.100247. PMID: 37456535; PMCID: PMC10344932.

Gupta, Namrata. (2015). Rethinking the Relationship between Gender and Technology: A Study of the Indian Example. *Work Employment & Society* 29(4): 661–672.

Gupta, N., and Etzkowitz, H. (2021). Women Founders in a High-Tech Incubator: Negotiating Entrepreneurial Identity in the Indian Socio-Cultural Context. *International Journal of Gender and Entrepreneurship* 13(4): 353–372.

Gupta, Sangita Dutta, Raychaudhuri, Ajitava, and Haldar, Sushil Kr. (2015). Information Technology Sector in India and Gender Inclusivity. *Gender in Management: An International Journal* 30(2): 94–108.

Haas, L., and Hwang, C.P. (2007). Gender and Organizational Culture: Correlates of Companies' Responsiveness to Fathers in Sweden. *Gender & Society* 21(1): 52–79.

Hewitt, Aon. (2016). Gender Pay Gap: Is It Deliberate? *India Forbes,* July 21, 2016. http://www.forbesindia.com/blog/business-strategy/gender-pay-gap-is-it-deliberate/. Accessed on September 4, 2024.

Holt, H., and Lewis, S. (2011). 'You Can Stand on Your Head and Still End Up with Lower Pay': Gliding Segregation and Gendered Work Practices in Danish 'Family-friendly' Workplaces'. *Gender, Work and Organization* 18: e202–e221. Available at: https://doi.org/10.1111/j.1468-0432.2009.00501.x. Accessed on September 4, 2024.

Inc42 (2022). Unicorns of India. Decoding India's 100 Unicorns. Report 2022. May 4, 2022. Inc42 Media. Available at: https://issuu.com/inc42/docs/inc42_s_100_unicorns_report_2022. Accessed on September 4, 2024.

Inc42 (2023). Indian Tech Startup Funding Report 2023. Inc42 Media. Available at: https://inc42.com/reports/#reports. Accessed on September 4, 2024.

Jain, D. (2018). What Prevents Women from Working in India? *Mint*, March 8, 2018. Available at: https://www.livemint.com/Politics/jedAN6zwNy0V0eXmc2vMGM/What-prevents-women-from-working-in-India.html. Accessed on September 4, 2024.

Jain, N., and Mukherjee, S. (2010). The Perception of Glass Ceiling in Indian Organisations: An Exploratory Study. *South Asian Journal of Management* 17(1): 23–42.

Jain, N. (2024). Ethical AI - Women's Critical Role in Shaping the Future of Technology, NASSCOM community, September 2, 2024. Available at: https://community.nasscom.in/communities/women-leading-change/ethical-ai-womens-critical-role-shaping-future-technology. Last accessed on November 13, 2024.

Jaswal, M. (2024). India's More Than 6,000 Women-Led Startups Remain Unfunded: Tracxn. *Mint,* March 8, 2024. Available at: https://www.livemint.com/companies/start-ups/indias-more-than-6-000-women-led-startups-remain-unfunded-tracxn-11709885614935.html. Accessed on September 4, 2024.

John, P. (2016). Gendered Medicine: A Study Among Medical Educators In Maharashtra. *BMJ Global Health* 1: A37–A38. Available at: https://gh.bmj.com/content/1/Suppl_1/A37. Accessed on September 4, 2024.

Karan, A., Negandhi, H., Hussain, S., Zapata, T., Mairembam, D., Graeve, H.D., Buchan, J. & Zodpey, S. (2021). Size, Composition and Distribution of Health Workforce in India: Why, and Where to Invest? *Human Resources for Health* 19(39):1-14., Open Access. Published 22 March, 2021. Available at: https://human-resources-health.biomedcentral.com/articles/10.1186/s12960-021-00575-2. Accessed on September 4, 2024.

Kelkar, G., Shrestha, G., and Veena, N. (2002). IT Industry and Women's Agency: Explorations in Bangalore and Delhi, India. *Gender, Technology and Development* 6(1): 63–84.

Korreck, S. (2019). *Women Entrepreneurs in India: What Is Holding Them Back?* ORF Issue Brief No. 317, September 2019, Observer Research Foundation: New Delhi. Available at: https://www.orfonline.org/wp-content/uploads/2019/09/ORF_Issue_Brief_317_Women_Entrepreneurs.pdf._Accessed on September 4, 2024.

Kumar, N. (2001). Indian Software Development: International Perspective. *Economic & Political Weekly* 36(45): 4278–4290.

Kumari, J.S., Lakshmi, S.G., Chandra Shekharan, P.A. and Bindu, C. (2015). Work Place Issues and Challenges Faced by Women Doctors in Clinical departments. *IOSR Journal of Dental and Medical Sciences (IOSR-JDMS)* 14(11), Ver. V (Nov. 2015): 51–61.

Kundra, S., Sreen, N., and Dwivedi, R. (2023). Impact of Work from Home and Family Support on Indian Women's Work Productivity during COVID-19. *Vikalpa* 48(1): 39–53.

Kundu, Subhash C. (2003). Workforce Diversity Status: A Study of Employees' Reactions. *Industrial Management & Data Systems* 103(4): 215–226.

Lewis, S., J. Brannen, and A. Nilsen. (2009). Work, Families and Organisations in Transition. European Perspectives. London: Policy Press.

Lewis, Suzan, and Humbert, Anne Laure. (2010). Discourse or Reality?: "Work-Life Balance", Flexible Working Policies and the Gendered organization. *Equality, Diversity and Inclusion: An International Journal* 29(3): 239–254.

LinkedIn. (2024). Women in Leadership in Corporate India. Available at: https://thequantumhub.com/wp-content/uploads/2024/05/Women-in-Leadership_Linkedin_Final_p1-32_31May.pdf. Accessed on September 4, 2024.

Maji, S., and Dixit, S. (2020). Gendered Processes and Women's Stunted Career Growth: An Exploratory Study of Female Software Engineers. *The Qualitative Report* 25(8): 3067–3084.

Makers India Report (2020). State of Women in Tech Entrepreneurship in India. A Report by Makers India and Yourstory (Jan 2018-June 2020). Available at: https://yourstory.com/cs/uploads/Finalversion-MakersWomeninTechReport-1613978390144.pdf. Last accessed on November 14, 2024.

Mani, S. (2023). Slow Rise in Women Numbers in Top Indian IT firms. *Times of India,* May 27, 2023. Available at: https://timesofindia.indiatimes.com/city/bengaluru/slow-rise-in-women-nos-in-top-indian-it-firms/articleshow/100541136.cms?from=mdr. Accessed on September 4, 2024.

Mittal, R. (2021). Arts-A Subject Just for Girls? *Boyish.* 2021. Season-I. Available at: https://boyish.in/season-1/content/issue-2/. Accessed on September 4, 2024.

MIWE (2021). *The Mastercard Index of Women Entrepreneurs.* Report. March 2022. https://www.mastercard.com/news/media/phwevxcc/the-mastercard-index-of-women-entrepreneurs.pdf. Accessed on September 4, 2024.

MSI-IT (2016). Indian IT Sector. Monster Salary Index. Wage Indicator Foundation. Available at: https://media.monsterindia.com/logos/research_report/MSI_Report_Low_IT_2016_.pdf. Accessed on September 4, 2024.

MSI (2017). Monster Salary Index. Available at: https://www.foundit.in/media/logos/research_report/MSI_Full_Report_2016_July_2017.pdf. Accessed on September 4, 2024.

Nasscom-Zinnov (2022). *Rising Above Uncertainty: The 2022 Saga of Indian Tech Start-Ups.* Zinnov-Nasscom India Tech Start-Up Landscape Report 2022. Available at: https://zinnov.com/innovation/zinnov-nasscom-india-tech-start-up-landscape-report-2022/. Accessed on September 4, 2024.

Nath, G. (2000). Gently Shattering the Glass Ceiling: Experiences of Indian Women Managers. *Women in Management Review* 15(1): 44–52.

NER (2016). National Employability Report-Engineers. Aspiring Minds. Annual Report 2016. Available at: https://quantumuniversity.edu.in/qlib/document/pdf/National%20Employability%20Report%20-%20Engineers%20Annual%20Report%202016.pdf. Accessed on September 4, 2024.

NER (2019). *National Employability Report-Engineers.* Aspiring Minds. Available at: https://www.scribd.com/document/653368364/National-Employability-Report-Engineers-en-In. Accessed on September 4, 2024.

Nikore, M. (2022). How COVID-19 Deepened the Gender Fault Lines in India's LaborMarkets. *Economic and Political Weekly (Engage)* 57(51): 1–19. ISSN (Online): 2349–8846. Available at: https://www.epw.in/engage/article/how-covid-19-deepened-gender-fault-lines-indias. Accessed on September 4, 2024.

NIRF (2018). *National Institutional Ranking Framework.* MHRD. India Rankings 2018. Available at: https://www.nirfindia.org/Rankings/2018/Ranking2018.html. Accessed on September 4, 2024.

Niti Aayog (2015). *Report of the Expert Committee on Innovation and Entrepreneurship.* NitiAayog, GOI: New Delhi. Available at: https://www.academia.edu/23331636/Report_of_the_Expert_Committee_on_%20Innovation_and_Entrepreneurship?email_work_card5view-paper. Accessed on September 4, 2024.

Oxfam Report (2022). *India Discrimination Report 2022.* Oxfam India. Available at: https://www.oxfamindia.org/knowledgehub/workingpaper/india-discrimination-report-2022. Accessed on September 4, 2024.

Pal, J. (2010). *Of Mouse and Men: Computers and Geeks as Cinematic Icons in the Age of ICTD.* iConference 2010 papers (2010-02-03). Available at: https://www.ischool.berkeley.edu/sites/default/files/joyojeet_pal_of_mouse_and_men.pdf. Accessed on September 4, 2024.

Pallissery, F., Beerannavar, C.R., and Thomas, F. (2023). Women on the Board of Indian IT Companies: Are They Audible and Visible?. In M.A. Naser (ed), *New Approaches to CSR, Sustainability and Accountability, Volume IV. Accounting, Finance, Sustainability, Governance & Fraud: Theory and Application*, Springer: Singapore, 177–195. https://doi.org/10.1007/978-981-16-9499-8_10

Pandey, P., and Pansera, M. (2020). Bringing Laxmi and Saraswati Together: Nano-Scientists and Academic Entrepreneurship in India. *Technology in Society* 63(November 2020): Article 101440. Available at: https://www.sciencedirect.com/science/article/pii/S0160791X20302657. Accessed on September 4, 2024.

Parikh, P. P., and S. P. Sukhatme. (2004). 'Women Engineers in India'. *Economic & Political Weekly* 39 (2): 193–201.

Patel, R., and Parmentier, M.J.C. (2005). The Persistence of Traditional Gender Roles in the Information Technology Sector: A Study of Female Engineers in India. *Information Technologies and International Development* 2(3): 29–46.

PP-WTC report (2023). *Pink Collar Skilling: Unleashing the Women's Power in the Real Estate Sector. Primus Partners.* Available at: https://www.primuspartners.in/reports/pink-collar-skilling-unleashing-the-women-power-in-real-estate-sector. Last accessed on November 14, 2024.

PwC (2016). *Making Diversity Work. Key Trends and Practices in the Indian IT-BPM Industry.* NASSCOM. Available at: https://www.pwc.in/assets/pdfs/publications/2016/making-diversity-work-key-trends-and-practices-in-the-indian-it-bpm-industry.pdf. Accessed on September 4, 2024.

Radhakrishnan, S. (2008). Examining the 'Global' Indian Middle Class: Gender and Culture in the Silicon Valley/Bangalore Circuit. *Journal of Intercultural Studies* 29(1): 7–20.

Radhakrishnan, S. (2009). Professional Women, Good Families: Respectable Femininity and the Cultural Politics of a 'New' India. *Qualitative Sociology* 32(2): 195–212.

Ramakrishnan, Aditi, Dana Sambuco, and Reshma Jagsi. (2014). Women's Participation in the Medical Profession: Insights from Experiences in Japan, Scandinavia, Russia, and Eastern Europe. *Journal of Women's Health (Larchmt)* 23 (11): 927–934.doi:10.1089/jwh.2014.4736

Rao, M., Rao, K.D., Kumar, A.S., Chatterjee, M., and Sundararaman, T. (2011). Human Resources for Health in India. *Lancet* 377(9765): 587–598. https://www.thelancet.com/journals/lancet/article/PIIS0140-6736(10)61888-0/fulltext. Accessed on September 4, 2024.

Ridgeway, C.L., and Correll, S.J. (2004). Unpacking the Gender System: A Theoretical Perspective on Gender Beliefs and Social Relations. *Gender & Society* 18(4): 510–531.

Rincon, R., Korn, R.M., and Williams, J.C. (2019). *Examining Gender Bias in Engineering in India.* Paper presented at 2019 ASEE Annual Conference & Exposition, Tampa, Florida. https://doi.org/10.18260/1-2—32777

Samantroy, E., and Tomar, J.S. (2018). Women Entrepreneurship in India: Evidence from Economic Censuses. *Social Change* 48(2): 188–207.

Skillsoft Report (2022). *Skillsoft's 2022 Women In Tech Report. INDIA Region.* Available at: https://s3.us-east-1.amazonaws.com/skillsoft.com/prod/images/Skillsofts-Women-in-Tech-Report-India-Region-Final-Mar-8.pdf. Last accessed on November 14. 2024.

Shanker, Deepika. (2008). Gender Relations in IT Companies: An Indian Experience. *Gender, Technology and Development* 12(2): 185–207.

Sharma, S., and Kaur, R. (2019). Glass Ceiling for Women and Work Engagement: The Moderating Effect of Marital Status. *FIIB Business Review* 8(2): 132–146.

S&T Indicators (2023). *S&T Indicators Tables 2022–2023*. Research and Development Statistics. Ministry of Science and Technology: New Delhi. Available at: https://dst.gov.in/sites/default/files/S%26T%20INDICATORS%20TABLES%202023.pdf. Accessed on September 4, 2024.

Shrestha, G., Joshi, D., and Clément, F. (2019). Masculinities and Hydropower in India: A Feminist Political Ecology Perspective. *International Journal of the Commons* 13(1): 130–152.

Shukla, T., Chauhan, G.S., and Saumya, S. (2018). Traversing the Women Entrepreneurship in South Asia: A Journey of Indian Startups through Lucite Ceiling Phenomenon. *Journal of Enterprising Communities: People and Places in the Global Economy* 12(1): 50–66.

Singh, N. (2018). Women Replace Men as HR Heads in 60% Companies. *The Times of India*. December 25, 2018. Available at: https://timesofindia.indiatimes.com/articleshow/67237898.cms?utm_source=contentofinterest&utm_medium=text&utm_campaign=cppst. Accessed on September 4, 2024.

Skillsoft Report (2022). Skillsoft's 2022 Women In Tech Report. INDIA Region. Available at: https://s3.us-east-1.amazonaws.com/skillsoft.com/prod/images/Skillsofts-Women-in-Tech-Report-India-Region-Final-Mar-8.pdf. Last accessed on November 14. 2024.

Sood, Mamta, and Chadda, R.K. (2010). Women in Medicine. *Indian Journal of Gender Studies* 17(2): 277–285.

SWE (2024). *The Current Status and Path Forward of Women Engineers in India*. All Together. Society of Women Engineers. Available at: https://alltogether.swe.org/2024/06/inwed-women-engineers-in-india/. Accessed on September 4, 2024.

SWI (2023). *State of Working India 2023*. Social Identities and Labor Market Outcomes. Azim Premzi University. Available at: https://azimpremjiuniversity.edu.in/publications/2023/report/state-of-working-india-2023-social-identities-and-labour-market-outcomes. Accessed on September 4, 2024.

Thayyil, J., and Jeeja, M.C. (2013). Issues of Creating a New Cadre of Doctors for Rural India. *International Journal of Medicine and Public Health* 3(1): 8–11.

Thelwall, Mike, Bailey, Carol, Makita, Meiko, Sud, Pardeep, and Madalli, Devika P. (2019). Gender and Research Publishing in India: Uniformly High Inequality? *Journal of Informetrics* 13(1): 118–131.

UNESCO Science Report (2021). To Be Smart, The Digital Revolution Will Need to Be Inclusive: Excerpt from The UNESCO Science Report. Available at: https://unesdoc.unesco.org/ark:/48223/pf0000375429. Accessed on September 4, 2024.

Upadhye, Carol. (2006). Gender Issues in the Indian Software Outsourcing Industry. In Gurumurthy Anita, P. J. Singh, A. Mundkur and M. Swamy (eds), *Gender in the Information Society: Emerging Issues*, UNDP-APDIP: Bangkok, 74–84. Available at: https://www.unapcict.org/sites/default/files/2019-01/GenderIS.pdf. Accessed on September 4, 2024.

Varma, R. (2010). Computing Self-Efficiency among Women in India. *Journal of Women and Minorities in Science and Engineering* 16(3): 257–274.

Vyas, L. (2023). Experiences of Working Women in India under Family-Friendly Policies: Straight from the Horse's Mouth. *Journal of Asian Public Policy*. published online. https://doi.org/10.1080/17516234.2023.2191802 https://www.tandfonline.com/doi/abs/10.1080/17516234.2023.2191802). Accessed on September 4, 2024.

WE (2023). IT Companies Employed More Women in 2022–2023, with an Increase of 6.5%. *Women Entrepreneur. India*, Friday, June 2, 2023. Available at: https://www.womenentrepreneurindia.com/viewpoint/experts-column/it-companies-employed-more-women-in-2022-2023-with-an-increase-of-65-nwid-3431.html#:~:text=According%20to%20the%20study%20done,which%20was%205.9%20per%20cent. Accessed on September 4, 2024.

WEF (2020). *Global Gender Gap Report 2020*. WEF. Available at: https://www3.weforum.org/docs/WEF_GGGR_2020.pdf. Accessed on September 4, 2024.
WEF (2024). *Global Gender Gap Report 2024*. World Economic Forum. Available at: https://www3.weforum.org/docs/WEF_GGGR_2024.pdf. Accessed on September 4, 2024.
WIPO (2023). *Global Innovation Index 2023. India*. At:https://www.wipo.int/edocs/pubdocs/en/wipo_pub_2000_2022/in.pdf. Accessed on September 4, 2024.
WIPO-IP (2022). *World Intellectual Property Indicators 2022*. World Intellectual Property Organization. Available at: https://www.wipo.int/edocs/pubdocs/en/wipo-pub-941-2022-en-world-intellectual-property-indicators-2022.pdf. Accessed on September 4, 2024.
WISAT (2012). National Assessments on Gender Equality in the Knowledge Society Country Results: India. Nair, Sudha. WISAT. The Elsevier Foundation. At http://wisat.org/wp-content/uploads/National_Scorecard_India_Reduced.pdf. Accessed on September 4, 2024.
Women and IT Scorecard (2017): *Women and IT Scorecard – India, A survey of 55 firms*. February 2017. Report number: 1; The Open University Project: Gender, Skilled Migration and IT: A comparative study of India and the UK. By Parvati Raghuram, Clem Herman, Esther Ruiz Ben, Gunjan Sondhi. https://doi.org/10.13140/RG.2.2.10118.98882
Women and IT Scorecard (2018). *Women and IT Scorecard-India 2018*. NASSCOM and GSM-IT. The Open University, UK. Available at: https://www.researchgate.net/publication/328603083_Women_and_IT_Scorecard_-_India_2018. Accessed on September 4, 2024.
Woods, Mary N. (2017). *Women Architects in India. Histories of Practice in Mumbai and Delhi*. Routledge: Oxon and New York.
World Bank (2018). *Female Entrepreneurs: How and Why are They Different?* Jobs Working Paper, Issue no. 20. By Eliana Carranza, Chandra Dhakal and Inessa Love, International Bank for Reconstruction and Development/ World Bank: Washington, DC. Available at: https://documents1.worldbank.org/curated/en/400121542883319809/pdf/Female-Entrepreneurs-How-and-Why-are-They-Different.pdf. Accessed on September 4, 2024.

4

GENDER AND STEM RESEARCH

Part 1

Introduction

The gender gap between men and women as regards participation, top positions, and awards in STEM is a product of interrelated institutional, organizational, and socio-cultural factors. This chapter begins with the numbers indicating the gender gap in science in terms of participation and rewarding of 'merit'. It then analyzes the story behind the numbers that continues into the next chapter. This chapter is divided into two sections. The first section discusses the relationship between women's status and 'merit' in the international and national context. The second part analyzes the informal milieu of women researchers in the training phase, or the doctoral period. The gender biases encountered in academia in this phase mark the beginning of the disadvantages that accumulate in a woman's career.

Women in academic/research careers: representation and segregation

There has been an improvement in the status of women scientists over the last two decades. The proportion of women in R&D has increased from 10.2% at the beginning of the millennium (R&D Statistics 2000-2001) to 18.6% in 2021 (S&T Indicators 2023: Table 22). The proportion in many of the elite institutes has also been rising. For instance, the proportion in IIT Kharagpur has risen from about 10% in 2014, to about 14% in 2022; and IIT Delhi from 16.5% in 2015 to 19.2% in 2022.[1] Apart from the rise in women's participation, there has also been a growing representation in leadership positions and awards. While the first woman ever to head a Council for Scientific and Industrial Research (CSIR) lab was appointed in 2013,

DOI: 10.4324/9781003596660-5

TABLE 4.1 Women in scientific research in India

Women doctorates	In science 46%; engineering 29% [a]
Women's share in total R&D manpower	India: 18.6% (2021) [b]; World average: 29.3% (UIS 2019)
Women's share of the total R&D manpower in the higher education sector (2021)	India: 13% [b]
Proportion of women in R&D in Central R&D agencies (2021)	20% [b]
Proportion of women faculty	Older IITs: below 20%; IISc: about 8% [6]
Women in leadership positions	6% of women vice-chancellors in govt-run universities (IAP 2016); no female directors of IITs/IISc except at Zanzibar (IIT-Madras)
Awards	3.4% women awardees of Swarup Bhatnagar Award until 2021[7]
Fellowships	9.5% women as INSA fellows[8] and 1.7% as INAE fellows until 2021[9]
Women's share in R&D personnel in private sector (2021)	17.2% [b]
Women employed in the private sector as a proportion of total women in R&D (2021)	26.4% [b]
Female share of extramural grants	33% in 2013–2014; 25% in 2019–2020 [c]

Sources: a. AISHE (2021–2022); b. S&T Indicators 2022–2023; c. R&D Statistics at a Glance 2022–2023: Figure on 'Gender Participation in Extramural R&D Support by Central S&T Agencies'; 6,7,8,9: Please see endnotes.

subsequently more such headships have been awarded to women; and, at present the director general of CSIR (with a network of 37 labs across the country) is a woman. Women are being appointed as deans at the IITs, while their numbers were almost nil at the beginning of the millennium. The proportion of women fellows in the Indian Academy of Sciences has increased from 2.98% in 1995 to 9.8%.[2] Women scientists are increasingly receiving recognition through presidentship of academies, awards, and establishment of chairs.[3]

Nevertheless, the progress is excruciatingly slow (Table 4.1). Women's share in total R&D manpower (including the industrial sector), at 18.6% (S&T Indicators 2023: Table 22), is considerably lower than the world average of 29.3%; and 18.5% for South and West Asia (UIS 2019). Women's share in the higher education sector is a mere 4.9% of the total R&D manpower and about 13% of those in this sector (S&T Indicators 2023: Table 21, 22). Considering that the proportion of female doctorate holders in 'science' (including mathematics), 'engineering and technology' is around 38%

(about 46% and 29%, respectively: AISHE Report 2021–2022, Table 36), the proportion in research careers is extremely low. The significance of utilizing women's research potential needs to be seen in the overall Indian standing in global research.

Compared to the population of the country, India has a low research population. While the world average of researchers per million population was 1592 in 2018 (World Bank 2024b), India had only 262 in 2020 (S&T Indicators 2023: Table 29); although this figure is considerably higher than in the year 2000 (at 110 researchers/million). The world average expenditure on R&D as a proportion of the GDP was 2.62% in 2021 (World Bank 2024a); however, India spent only 0.64% between 2020 and 2021 (S&T Indicators 2023: Table 3). India's growth in research productivity is commendable but it lags behind in impact. India's growth rate of scientific publications (10.73%) has been much higher than the world average (3.83%) during 2008–2018, and India's share in the publication output is the third highest at 5.31%. However, the latter is much lower compared to about 20% for China (highest).[4] In terms of impact, India falls further, at the fifth position among the nations with top 1% of cited articles.[5] A full utilization of women's research potential, while being an issue of human rights, is also essential to enhance India's S&T capability.

That women experience inequality is visible in the form of vertical segregation. The ratio of women declines with the rise in academic positions in STEM institutes (Kumar 2001). More recent figures, but for all fields combined, show that the proportion of women falls from about 44% of Lecturers/Assistant Professors to 38% of Readers/Associate to 29% of Professors (AISHE Reports 2021–2022, Table 51). Of the top 100 universities under the National Institutional Ranking Framework (NIRF) 2022, a mere 11 are led by female Vice-Chancellors; and, in stark contrast, almost half of the top 100 colleges have women principals and also a significantly large group of female faculty (Jolly 2022). The proportion of women at the older and more prestigious IITs remains below 20%, and about 8% base rate at IISc.[6]

Women scientists lag behind in leadership positions and recognition. Women are seldom appointed as chairpersons of institute-level committees or as the heads of departments; there are only 6% of women Vice-Chancellors of government-run universities, and these are mostly women universities (IAP 2016). There have been no women directors of the IITs in India. A female director was appointed at IIT-Madras's new Zanzibar campus in July 2023. This appointment is overseas and was made by IIT Madras, and not by the usual procedure requiring the final approval of the President of India. More than ten new directors appointed in 2022 have all been male (Jolly 2022). There have been no women chairpersons of examination bodies for entrance exams of bachelor's (or the JEE Advanced) and master's (or GATE) of the IITs. Women scientists receive fewer awards. Only 19 women (3.4% of the total awardees) until 2021 have received the coveted Shanti Swarup Bhatnagar (SSB) Award instituted in 1958.[7] Of the 13 awardees, only one woman scientist received the recently instituted Vigyan Shri in 2024. The women fellows of INSA were 9.5% in 2021[8] and of INAE are 1.7% (15/878).[9] Men continue to occupy leadership positions in the academies in overwhelming numbers (UNESCO 2022).

Further, there is horizontal segregation with the proportion of women varying for different fields. Women constitute about 57.4% of the total doctorates awarded in bio-science, while only about 39% in physics (AISHE Reports 2021–2022: Table-36). However, compared to the proportion of doctorates, the proportion of women scientists employed is low for both fields. For instance, there are about 23% of women scientists in research institutes that mainly focus on biological sciences (Swarup and Dey 2019); in physics, the proportion is about 16% of the faculty/scientists at the physics departments of academic institutes and physics research laboratories (Dabas 2021). While there are about 29% of female doctorates in engineering (AISHE Reports 2021–2022: Table 36), the proportion of women as faculty in engineering is estimated to be 9.2%.[10] Certain departments of engineering, particularly, mechanical engineering, might have very few women at the older IITs. Counting manually from their websites, there are 2/47 in IIT Kharagpur and 5/52 in IIT Delhi (as of 30 May 2023). However, even this is an improvement over the last two decades. The women faculty in mechanical engineering departments were non-existent at these two older IITs in a study conducted at the beginning of the millennium (Gupta and Sharma 2003). An increase in the numbers, even if modest, is important as it helps to bring support and changes in attitudes (Etzkowitz et al. 2000). Thus, there is change, but it is quite sluggish. Table 4.1 encapsulates the current status of women in scientific research.

Does a low proportion of leadership positions and awards among women scientists imply that there are few meritorious women, or is there a relationship between gender and merit?

Section 1: gender and 'merit '— the institutional context

'Universalism' in science, or that the scientific contributions are judged objectively and careers are open to talent with evaluations based on merit, is one of the important norms in the Mertonian ethos of science (Merton 1942). However, studies amply demonstrate 'particularism' in science. Researchers have examined different elements of academic excellence, such as journal rankings, citation indexes, peer review systems, and grant applications, in order to demonstrate that hegemonic structures of inequality based on gender, race, and class are being reproduced in judging merit (e.g., Adler and Harzing 2009; Laudel 2006; European Commission 2019). Papers published by women receive systematically fewer citations than papers written by men (Teich et al. 2022). Studies have found that women scientists are less likely to be selected for prominent awards reflecting the 'Matilda effect' or bias against acknowledging achievements of women scientists whose work gets attributed to their male colleagues (Lincoln et al. 2012). The case of Rosalind Franklin, whose research on DNA was not duly credited by the male Nobel prize winners, is well-known. The male researchers have the 'first-mover advantage' as found by Kong et al. (2022); that is, among similar papers, those authored by men were published before those by women authors. As a result, men can establish themselves in the field earlier than women do, which allows them to grow their network and ultimately get more recognition.

Thus, productivity/citations as the criteria of 'merit' are essentially flawed. Nevertheless, they continue to be relevant in judging 'merit'. This makes it imperative to examine the relationship between Indian women's lack of status in science and their research publications/citations.

Productivity/citations

The studies in India show varying results regarding whether women's publications lag behind men's. In the West, there is vast literature documenting that the measured productivity is lower for female than for male scientists, almost in any discipline and whatever the productivity measure considered (e.g., Lee and Bozeman 2005).

However, in India, many studies indicate that there is no significant gap in research productivity. Gupta, Kumar, and Aggarwal (1999) found that the average productivity of Indian female and male scientists in the physical sciences, biology, and engineering sciences was not significantly different. Hasan et al. (2012), analyzing SCI publications of research scholars from CSIR labs between 2004 and 2009, concluded that the research output of female research scholars is at par with male research scholars and commensurate with their numerical strength in academic and R&D institutions. Lemoine's (1992) study of CSIR scientists showed that the proportion of less productive scientists was slightly greater for men than for women. In an analysis of publications of scientists at Bhabha Atomic Research Centre (BARC) for the period 2006–2010, while women formed 16% of the total number of scientists at BARC, they contributed to about 36% of the technical journal publications. (Upadhye et al. 2014).

However, some Indian studies have reported a lower research productivity rate for women compared to men, less female first authors and lower collaboration. Bal (2005) examined 669 publications indexed by PubMed during January 1994–April 2004 to study the scientific productivity of female biologists and found that quality of productivity is not commensurate with their proportion in employment and attributed this to women-specific environmental factors. Using the publication and citation data in the 12 sub-disciplines of life sciences in journals indexed by the Web of Science (WoS) database for 2008–2009, Garg and Kumar (2014) also found that women publish less in this field and publish in low-impact factor and domestic journals. In agricultural and social sciences, women are more likely to write single-authored articles and have fewer collaborators than men, although they are comparable to men in terms of citation indices and publish in higher-ranked journals more frequently than men (Bansal et al. 2022). Thelwall et al. (2019) found that compared to the US, India has a much lower share of female-first authors. Dentistry, economics, and mathematics are all more female in India, but veterinary is much less female than in the US. These differences, according to the authors, are probably an extension of subject choices at the tertiary education level.

Whether women scientists publish less or more than men does not seem to be important to women's position in science. The fact remains that neither a strong female doctoral pipeline nor comparable scientific productivity is able to influence the underrepresentation of women at the top ranks of academic/research institutes. This is confirmed through studies elsewhere also (e.g., D'Amico et al. 2011). Women scientists have a lower status in India despite the evidence of comparable productivity in several S&T fields. Kumar (2001) found no significant difference in the research productivity of men and women scientists, yet, there were significant differences in academic ranks of men and women, with women at lower ranks. 'Universalism' falters in science, and women are less able to translate their productivity into resources and recognition (Long and Fox 1995). A lower publication rate among women scientists in the West, as several studies show, is a product of several institutional and individual-level reasons (e.g., Fox 1983; Zainab 1999) and is both the cause and effect of gender disparity in career attainments in science (Fox 2006: 23). Thus, as Fox (2004) states, a publication gap is not a significant factor in women's lower positions in science.

Gender-based citation differences are particularly marked in India. In a study of gender differences in research impact through Scopus citations and Mendeley readers in five countries, including India, for articles published in 2014, Thelwall (2018) found that female-authored research is less cited in India compared to the US, UK, and Spain, and has the least number of Mendeley readers (suggesting a low audience for research) among the countries surveyed. Lower citations for female-authored papers were also found in another study for articles published over a ten-year period from 2008–2017 (Paswan and Singh 2020). There could be various reasons for this including socio-cultural biases, lack of resources, of mobility to attend prestigious institutes, and lack of networks among women, older scientists (male mostly) producing more visible research, and women in lower citation specialisms (Thelwall 2018; Gupta and Sharma 2002).

In India, productivity has been found related to age but not to marital status. A higher age is associated with greater funding, greater collaborations, wider networks, and more students, which lead to higher research productivity. A baseline study (GATI Report 2022) by DST and the British Council in 2021 to study the status of women in STEM institutes for the Gender Advancement for Transforming Institutions (GATI) project found that, while the marital status of women had no significant impact on the number of publications, age, on the other hand, had a positive and significant effect. As the age increases, the number of publications also increases i.e., those above the age of 46 years accounted for 52.2% of the publications. Gupta and Sharma (2002) found that women scientists are most productive researchers in their 50s. In a study of speech and hearing faculty (Subramanian 2017), the peak productivity was observed for the female teaching faculty at >55 years (with a rising trend in the 50s), while the peak productivity for the male teaching faculty was observed at <30 years (although an increase into the 50s is steady). Similar findings are reported for agricultural and social sciences: that although women have fewer numbers of publications in their early careers, they catch up later but lag behind men in seniority (Bansal et al. 2022). A higher productivity rate for those in their 50s could be due to lower childcare commitments and greater experience and networks. On the other hand, gendered networks and commitments seem to constrain early peak productivity that might affect a rise in the organizational ladder.

Indian research: international status and gender

As noted previously, although India's research productivity is growing at the fastest pace in the world, it is third in terms of output and fifth in highly cited nations, it continues to be at the 'periphery'. Its citation impact was 25% less than the world average and accounted for only 3.4% of the world citations and 3.9% of the international collaboration output in 2013; engineering, which is the most impactful field, is cited 6% less than the world average between 2009–2013 (Elsevier 2016). Thus, there seems to be a stratification in science at the international level.

Despite the emergence of a more multipolar world in terms of scientific output, including China and India, there continues to be a hierarchy in the global system of science, with Euro-American domination in the content of science, leading institutions, publishing regime, and language (Marginson and Xu 2021). Non-white scholars continue to be underrepresented in publication rates, citation rates, and editorial positions (Chakravartty et al. 2018). Non-white scientists not only appear on fewer editorial boards, but also endure longer waiting times between the submission and acceptance of their manuscripts; and upon publication, their papers receive fewer citations than would be expected based on textual similarity (Liu et al. 2023). The elite scientists of Asia pursuing state-of-the-art research are usually trained in the West, and on their return to Asian institutions, suffer disappointment partly due to a loss of status owing to the structural biases in the global scientific field that privileges research produced by Western research organizations over work from elsewhere (Paul 2021).

The Western white hegemony in scientific research is likely to hurt women more than men. Intersection of gender and race could lead to novel contributions of women being discounted (Hofstra et al. 2020). The center-periphery dichotomy is essentially a hierarchy, and women scientists of the periphery would lie at the bottom. Women in developing countries are less likely than men to have trained abroad and to travel, leading to 'research localism' (p. 472), restricted professional networks, and lesser proficiency in English (Campion and Shrum 2004).

Interestingly, women researchers in India do not seem to lag behind their male colleagues in international collaborations. A study found that female authored papers have a higher number of internationally collaborated papers than men in the period between 2008 and 2017, except for two years (Paswan and Singh 2020). Women scientists value their international contacts more highly than men, possibly due to the knowledge exchanged and the style of interaction (Palackal et al. 2007).

Being an internationalist, that is, maintaining contacts and collaborating with scientists in the West, enhances the reputation of the scientist (Subramanyan 1998). However, this is insufficient and recognition is linked to the local science structures and processes. As Subramanian (2015) noted, a scientist is recognized if someone from her field of specialization within the country knows someone abroad who thinks that her work deserves recognition. To be known within the country among the scientists who matter, one has to lobby, which is difficult for a woman, particularly if competing against a man.

Thus, women scientists in a developing country like India are victims of both: a gendered dimension of merit, which is universal; and of an international science structure that peripheralizes the science of developing countries. Further, one needs to take into account the notion of 'merit' as defined and rewarded by the national scientific community and scientific organizations in India, which construct a gendered 'merit' in the Indian context.

Gendered notion of 'merit': the Indian context

The definition of 'merit' manifests itself in the discourse among scientists. Scientists in India equate 'merit' with the talent for doing science, which is believed to be inherent in a person and will become apparent irrespective of external conditions (Subramanian 2007). Merit in India, as in the West, is believed to manifest early. This belief is exemplified in India in a low age of recruitment (maximum being 35 years). Merit at the entry level faculty positions in academia is judged in terms of linear progression of obtaining academic credentials with minimum breaks. This disadvantages women for whom the early 30s coincides with marriage and childbearing. Saha and Ray (2015) found that Indian academic scientists tend to become not only more active, but also more productive in research over their lifetime. Hence, raising age limits for recognitions while benefitting women will help male researchers, also. In view of this, new awards (Vigyan Ratna and Vigyan Shastri) with no age limits, established in the year 2023, is a positive development.

To be considered 'meritorious' requires winning awards/fellowships that usually call for networking with the top scientists in the field. Fellowships of national academies require that the scientist be nominated by existing fellows, and the application is then screened by sectional committees. It is an unwritten fact that knowing the decision-makers and members of various committees helps in a nomination and then a selection (Bal 2005; Gupta 2007b). Given the lack of social capital, women lose out in terms of recognition. Lack of recognition excludes women from powerful positions in decision-making bodies of the government. Absence of an open democratic approach in science institutions vests power in a few, creating a circulation of the same scientific elite in such bodies. Thus, such institutions cannot be 'universalistic' in rewarding merit (Subramanian 2007).

Absence of women in the decision-making bodies on the one hand and limitations in networking on the other, impacts research funding from external agencies. Research grants often require lobbying with the officials in funding agencies in Delhi when their proposals are evaluated by those agencies. Women usually engage in formal communication in such situations as informal interaction, which is more helpful, is limited for women both due to inappropriateness of interacting with men after official hours and the perception of manipulation as unbecoming of women (Subrahmanyan 2009). Obtaining funding and other opportunities in science hinges considerably on the review panel members being familiar with the candidate; research council meetings usually have the same set of members, implying that a small proportion of powerful scientists (usually male) control scientific resources (Dogra and Jayaraj 2023). The percentage of women receiving extramural grants for R&D by the central R&D agencies has increased to 25% in 2019–2020 from 13% in 2000–2001 due to various initiatives of the government (R&D Statistics at a Glance 2022–2023: 9). DST and DBT are two main players in extramural grants. However, the gender gap in the extramural grants has been increasing, with the female share in extramural

R&D support by central and S&T agencies declining from 33% in 2013–2014 to 25% in 2019–2020, implying an increase in the male share from 67% to 75% (R&D Statistics at a Glance 2022–2023: Figure on 'Gender Participation in Extramural R&D Support by Central S&T Agencies').

The organizational structure involved in practicing science suffers from various limitations that affect women more than men. One of the major limitations is a steep hierarchy of institutes in terms of resources, prestige, and quality of research.

Research and hierarchy of institutions

Despite the fact that India produces the third largest number of doctorates in science and engineering in the world after the US and China (R&D Statistics at a Glance 2022–2023), there are considerable faculty shortages in the country in most central universities and elite academic institutes, including the IITs that have about 40% unfilled teaching posts in 2022 as informed by the government in Parliament.[11] In the newly opened universities, it is as high as 80% (Sharma 2018). The reasons are due to remote locations of some universities and a shortage of funds. The higher education sector accounts for only 8.8% of the gross expenditure on R&D (R&D Statistics at a Glance 2022–2023). Since they are cash-strapped, universities prefer ad-hoc/temporary teachers as less salaries need to be paid; private universities/institutes tend to employ the bare minimum number of teachers to maximize profits; and elite technical institutes such as the IITs face shortages due to a lack of availability of qualified academics, particularly for the new courses.[12] The shortage of faculty exists along with a surplus of doctorates, which is compounded by a lack of industrial R&D, and, therefore, a low capacity to absorb PhDs (Pradeep 2018). In all of this, there appears to be a huge wastage of doctorates in the country both as underemployed and unemployed.

While lack of funds with the universities affects the numbers hired and the position offered (contractual/permanent), poor quality of doctoral programs renders the doctorates unemployable by well-known academic/research institutions. According to Krishna (2014), just about 10–12% of the universities and colleges are research intensive; the rest, nearly 88% of the universities, are only teaching institutions. Among the engineering and technological institutes, the IITs/IISc lead in research quality and quantity, and NITs are lower performers compared to the older IITs and some of the universities (Pratap and Gupta 2009). The most productive *private* institutes match with the IITS/IISc/NITs and central universities in the rate of growth of research output but lag in quality, although ahead of the NITs. Expenditure per faculty in private institutes is almost half of that in the national institutes, suggesting that resources are needed to scale up the research performance in private institutes (Banshal, Singh and Mayr 2019).

In this system, gender inequality is manifested through lower career outcomes for women than men. The proportion of women declines with the rise in the institutional hierarchy. Overall, the proportion of female teachers is about 38% of the total teachers at the university level and 44% at the college level (AISHE Reports 2021–2022: 38). There are a higher number of women teaching science and mathematics in colleges than universities (IAP 2016). The proportion of women faculty is higher in private engineering institutes than in the prestigious national institutes; for instance, among the top 20 engineering institutes, women as a proportion of the faculty was about 35% in Vellore Institute of Technology and 30% in Thapar Institute of Engineering and Technology (NIRF 2018). A high proportion of female faculty in private institutes was also noted in UNESCO's study for Manipal (UNESCO 2022).

The hierarchy of institutes maps to the gender hierarchy at the doctoral level of education. Women account for a much higher proportion of doctoral students (47.6%) in state public universities than in the Institutes of National Importance (32.5%: AISHE Reports 2021–2022, Table 50), owing probably to the family pressure of studying nearby. While this proportion includes all branches of learning, the proportion should roughly apply to those in S&T, also.

In many state universities, there is a lack of facilities and of mechanisms to share expensive resources/materials/machines with other NITs or the nearby IITs. At MDS University, Rajasthan, for instance, the research departments lack funding, and all but two of them cannot even afford to run undergraduate courses. The researchers are offered little help by the system to patent their work, to reach out to industries, or to collaborate with other research groups.[13] Geographically, research in the northeastern states is additionally hampered by a lack of infrastructure, poor governance, and lack of adequate manpower as most students move to metro or big cities (Paul 2021).

Lack of facilities exists for men also in such universities. However, discriminatory structures and lack of funding in state universities enhance the gendered professional outcomes (Kurup et al. 2017). Further, given married women's lack of geographical mobility, it is likely that the talented women are unable to move to better universities and institutes and to fully utilize their potential. On the other hand, marriage might require relocation that involves recruitment hassles at the new location. Often the state-funded universities and research labs in India follow the 'son of the soil' policy, recruiting a majority of the personnel from within the state. A woman moving outside the home-state due to marriage might find it difficult to find a job. Hence, a scientist recommended 25% seats for men and women in such institutes for meritorious candidates from other states (Amarnath 2011).

Despite the fund shortage and struggle to obtain lab materials and equipment, there are stories of high-quality research by women scientists in non-elite institutes. As, for instance, a known pteridologist in a Mangalore private institute, Smitha Hegde, with expertise on ferns of the Western Ghats, has been enlisted by the government for her help in controlling superweeds. However, her story also has another dimension, that of female doctoral students. According to Jayaraj (2016), due to a severe fund crunch, there were no fellowships for the PhD students and Dr Hegde barely scraped money from her projects for research. Yet, her lab had several female doctoral students, mostly married, willing to work with her without fellowship, as due to their husband's earnings money was not a big issue. Many such female students also dropped out from careers after their PhD due to domestic roles, though they might have taken up teaching later. It appears that gendered pressures on men to support themselves deter them from taking up doctoral studies in private institutes (suffering from fund shortage). For women, gendered domestic roles allow them to obtain degrees from anywhere, even without a research stipend. However, they might end up sacrificing their future research career.

In India, more than half of the R&D personnel are employed in the government sector (that includes the institutional sector, public sector industries, and elite educational institutes funded by the government). The central government incurs about 84% of the R&D expenditure on 12 major scientific agencies; women, however, constitute only about 20% (11,339/57,137) of the personnel in R&D in these agencies (R&D Statistics at a Glance; S&T Indicators 2022–2023).

The proportion of women scientists as project heads and division heads is higher (at 30%) in the agencies that do not require application for grants, such as Department of Space (DOS), Department of Atomic Energy (DAE), Defense Research and Development Organization (DRDO) or the Indian Space Research Organization (ISRO) than in others such as Department of Science and Technology (DST) and Department of Bio-Technology (DBT) (Godbole and Ramaswamy 2015). This suggests that gendered funding plays a major role in creating a lower status for women, and therefore, absence of the application requirement enables funding and a rise in the position for women. The project director of the 'Agni-4' missile in DRDO was a woman; and women scientists' roles in the launch of the Chandrayaan mission (Indian Lunar Exploration Programme by ISRO) are well known. DRDO had its first woman director general (R&D-Electronic and Communication Systems) in 2015. In 2018, DRDO's four director general positions (a dozen in DRDO in total), five directors of laboratories (a network of more than 52 labs) and a couple of project director positions were held by women (Somasekhar 2018).

There are two aspects of this rise of women in powerful positions in some organizations and labs: one, though the proportion of women scientists is rising, it still remains low. For instance, DOS has 15.7% of women as science/technical staff (DOS Annual Report 2021–2022) and DRDO has about 15% (Deccan Chronicle 2017); two, in most labs and in academia, the proportion of women remains low at higher levels and in leadership positions. In DOS, women constitute only 4.9% of the total in Group C scientific/technical staff, while constituting 19% in the lower Group A (DOS Report 2021–2022). Women in the space mission 'Mangalyaan' or the Mars Orbiter Mission, have been a subject of much media attention. The status of women in space organizations is discussed next.

Women in Mars mission

The media representation of women of the Mars Orbiter Mission (MOM), or the 'rocket women', seems to contradict the stereotype of a typical scientist as a male and obsessed about science. At the same time, it also legitimizes state science as a progressive project (Arora 2022). However, similar to academia (Gupta and Sharma 2002), there is gendered segregation in informal spaces in space organizations, expectation to follow the 'male model' of devotion of long hours, lack of women in leadership positions, and, yet, denial by the scientists of any gender inequality at the space organization (Arora 2022). Arora (2022) found that the idea of non-hierarchy, openness, learning, and harmony as values were cherished by the organization surveyed and by its members, that such ideas were often associated as forming the bedrock of 'good science'. Yet, the nexus of state and science, visible in spatial design with elaborate security concerns, overlaid by multiple hierarchies, seemed to contradict the claim of 'openness'/non-hierarchy.

Women in MOM were multitasking, exerting both at work and at home, leading to physical exhaustion (Vaid 2019). The family support through the tough months of the mission (frequent travel, long hours, odd working hours) seemed crucial. Vaid (2019) found that the ISRO centers have a number of husband-wife teams and this aids women's careers in science. This also implies that the workplace demands become more manageable when the organizations do not create a 'two body problem'. Further, women with spouses in different occupations found it difficult to manage the dual burden, which suggests that the manageableness in the case of husband-wife teams in the centers is due to logistics convenient for the couple and not due to workplace friendliness.

The private industrial sector

The private research sector is a relatively new phenomenon in India, which received considerable impetus with liberalization of the economy in the

1990s. The share of the private sector industry in gross R&D expenditure was 36.4% in 2020–2021, lower than most advanced and emerging economies with more than 50% share of the private sector (R&D Statistics at a glance 2022–2023). The sector in 2021 employed 28.6% of the total R&D personnel in the country, of which 17.2% were women; and of the total women engaged in R&D activities, about 26.4% were in the private sector (S&T Indicators 2022–2023: Table 21, 22).

Although one-fourth of women researchers are in the private sector, there is a paucity of studies on such women. A study (Gupta 2017) of women scientists was conducted at two reputed Indian pharmaceutical private sector companies with a global clientele (referred to as Private1 and Private2 to honor the confidentiality contract). At the time of the survey, these companies were among the top ten research labs in patenting activity and were located in metropolitan cities. The pharmaceutical industry is one of the few industries in India that has demonstrated R&D capabilities in the private sector. It has been a rapidly advancing sector in the country, both in production and technological capabilities (Krishna and Chandra 2009). These R&D labs are elite in terms of their reputation as the best private research labs in the country, employment of a highly educated workforce, high salaries, plush organizational setting, and international reach. Most employees were Hindu upper caste, as in other studies (e.g., Gupta 2016). The average age of the researchers was less than 34.91 years (Gupta 2017), compared to those in the CSIR labs, for instance, which was 47.64 years (Gupta 2016). This indicates a younger population in private R&D, which might be due to a late beginning of research in the sector, mainly at the turn of this millennium.

The percentage of women in these labs varied from 6–14% in their various R&D units, suggesting representation of women similar to those in the government sector and indicating a severe lack of critical mass. Similarly, men occupied most of the leadership positions, with women as managers in scientific teams being 5% to 30% in these units, and those at higher management positions as miniscule or none (Gupta 2017). Although scientists in the private sector identify themselves more as employees of the company rather than with the professional community of scientists, various issues facing women in these labs resemble those faced by the women scientists in academia and the public sector. These are discussed in Chapter 5.

Does gender impact production of knowledge?

Scientists regard production of scientific knowledge through the scientific method as unbiased and uninfluenced by socio-cultural norms. However, research suggests that the issue of gender enters production of scientific knowledge in various ways. Science, Technology, and Society (STS) studies emphasize that knowledge (or the facts) is a social construction. Knorr-Cetina (1981) and others helped to document minute social interactions in laboratories and how these interactions form part of the process of invention and discovery. Gender assumptions may enter scientific research inadvertently through unexamined metaphors. For instance, Evelyn Fox Keller (2001) found that gendered expectations led to an assumption of sperm as active and ovum as passive, and the shift in recognition to both gametes as active helped to shape the thinking and experiments of scientists.

There is a strong relationship between the characteristics of scientists, the topics they choose, and the science they produce. Research shows homophily between social identities and topic, suggesting that diversity is not only related to social justice, but also to an expansion of the knowledge base (Kozlowski et al. 2022).

Generally, the belief of women scientists in India is that the conduct of science is essentially value-neutral, particularly, in how one interprets data (Chadha, Krishna, and Tripathi 2017). Yet, the question is: what is the data collected and how it is collected and interpreted. Vaidya (a prominent female neuroscientist and behavioral biologist) questions why male rats are most often studied in neuroscience and that the study of female rats might be unimportant or important, but unless experiments are done, one does not know (Chadha, Krishna, and Tripathi 2017).

The view that science is neutral and is just a discovery of facts about nature is denied by Vandana Shiva (2016), arguing that science and development are patriarchal projects. Shah, through her discussion of the reproductive cycle and technological interventions related to it for contraception or infertility assistance, contends that 'value-free' science reinforces the status quo, and the technology ends up reducing choices for a woman (Shah 2017).

There is a belief that failing to consider gender and sex in scientific research is limiting the benefits of science (Nature 2020) and that 'gendered innovations', that is, integrating sex, gender, and intersectional analysis at every stage of research could profoundly enhance human knowledge in the fields of science, medicine, and engineering (Schiebinger et al. 2011–2021). These researchers point out that gender bias also leads to missed market opportunities. In engineering, for example, considering short people as 'out-of-position' people leads to greater injury among women in automobile accidents.

There is no study on gendered innovations in the Indian context. However, a growing number of women engineers, not necessarily from the IITs, are contributing to the product and tech inventions that are friendlier. Gonslaves, former director of IIT Mandi, cites examples of women engineers designing products and services that make life easier for people. DesiCrew by Saloni Malhotra (an engineer from Pune) provides online IT services from small towns and villages (Gonslaves 2023); women engineers are devising platforms for the benefit of women, for instance, Prachi Gupta co-founding LinkedIn's Women in Technology.[14]

Production of knowledge is linked to the status of women scientists. For instance, a scientific fact is not recognized as a 'fact' unless it is so recognized by one's peers, through publications and citations. Yet, as seen earlier in this chapter, there are gender biases in these processes. Further, recognition could be delayed, as in the case of the research of Barbara McClintock, whose results were accepted as 'facts' 40 years after she completed her experiments. Thus, a lack of 'universalism' in science constructs masculinity in the production of knowledge itself.

Further, in every national socio-cultural context, gender inequality in scientific research careers is a product of social interactions, norms, processes, and behaviors in science organizations. The issue of how gender inequality is constructed in research careers in the Indian context is discussed in Section 2 on the doctoral phase, and followed through in a discussion on scientific research careers in the next chapter.

Section 2: the doctoral phase — accumulation of disadvantages

The doctoral period is a transition phase, from education to career. Most often, women accumulate a series of disadvantages in this phase, that create a deficit of 'career capital', thus constructing gender differences in careers. These disadvantages are a product of an interplay of an informal milieu of the institutes and socio-cultural norms in the life of a female doctoral student. This interplay has been dynamic with changes in socio-cultural norms.

In the first half of the twentieth century, women scientists had to grapple with intense patriarchal traditions in society and in their labs while pursuing doctoral studies. Historically, women were not expected to seek employment after education. This led to their exclusion from science and mathematics in the first half of the twentieth century. It is, therefore, remarkable that a few women scientists emerged in this period. These include Janaki Ammal in plant sciences (DSc in 1931 from University of Michigan) from Kerala; Asima Chatterjee in organic chemistry, who was the first woman scientist to be awarded DSc by an Indian university (in 1944, Calcutta University); and Anandi Bai Joshi who was the first woman doctor (medical degree from the US) at the end of the nineteenth century (Godbole and Ramaswamy 2008). They belonged to middle-class families and were encouraged by their fathers or their husbands. According to Sur (2009), all three women in CV Raman's lab (Nobel laureate in physics in 1930) at IISc Bangalore, Lalitha Chandrashekhar, Anna Mani, and Sunanda Bai, belonged to middle or upper-class families during a time in history that valued education of women. Yet, the supportive men were unable to counter their own prejudices at times. For instance, Anandi Bai, whose husband supported her initially, reportedly said (Thakkar 2008: 15):

> He [Anandi's husband] kept on taunting me [Anandi] that I was a free bird in a foreign land and that I had probably forgotten my 'poor, uncivilized and incapable' husband who wasn't as 'great' as me. On seeing an innocent photograph I had sent him, he made a remark that I appeared to have forgotten my tradition and culture as my *pallu* [loose end of a saree] was askew.

The traditional socio-cultural norms were reproduced in structures, processes, and interactions in scientific research in the first half of the twentieth century. According to Sur (2009), the norm of gender segregation was strictly followed by Raman (a Nobel laureate in physics) within his lab who frowned upon any interaction between men and women. Conditions of work were particularly difficult due to such norms, which also constricted physical spaces available to the women working in Raman's lab. A 'heightened visibility' (Kanter 1977) due to minority status of women scientists brought an intense scrutiny of their actions. There were microaggressions, for instance, when male colleagues questioned their academic credentials or ridiculed them in the guise of humor (such as giving nicknames to every woman student). Apart from the gendered interactions, the scientific institutions and their processes 'perpetuated their own gender biases' (Sur 2009: 119). Anna Mani and Sunanda Bai were denied doctoral degrees on flimsy pretexts, though they truly deserved it.

There are considerable differences in the socio-cultural milieu since then. This is evidenced in the high proportion of women entering STEM fields. There has been a rise in the proportion of women doctorates enrolled *in the institutes of national importance*, since 2003–2004 from about 11–17% (Gupta 2007b) to about 32% in the past two decades (AISHE Reports 2021–2022: Table 50). Doctorates awarded to women in science have risen in numbers and proportion (Table 4.2). The number of women receiving doctorates in engineering and medicine has doubled within a span of six years (Table 4.2).

TABLE 4.2 Comparison of number of doctoral degrees awarded in STEM to men and women in 2015–2016 and 2021–2022

Year	*Science*		*Engineering and technology*		*Medical sciences*	
	Female (percentage of total)	*Male*	*Female (percentage of total)*	*Male*	*Female (percentage of total)*	*Male*
2015–2016	2,575 (39.0%)	4,032	744 (26.7%)	2,041	343 (28.0%)	883
2021–2022	3,431 (46.3%)	3,977	1,824 (29.1%)	4,446	813 (39.2%)	1,260

Source: AISHE Reports 2015–2016; 2021–2022: Table-36

TABLE 4.3 Comparison of number of doctoral degrees awarded in PCM to men and women in 2015–2016 and 2021–2022

Doctorate degrees awarded	*2015–2016*		*2021–2022*	
	Female (percentage of total)	*Male*	*Female (percentage of total)*	*Male*
Physics	298 (33.2%)	598	514 (39.4%)	789
Chemistry	426 (32.9%)	868	754 (40.4%)	1,106
Mathematics	238 (40.5%)	350	167 (51.5%)	157

Source: AISHE Reports 2015–2016; 2021–2022: Table-36

The reasons for such a high proportion of women doctorates include, among others, women's education becoming increasingly important among the middle class, and the growth in the number of colleges that create respectable undergraduate teaching positions. However, a decline in the *number* of men receiving doctorates in science (Table 4.2) also points to a lack of attractiveness of the field for men. However, this trend does not hold true for physics and chemistry at the doctoral level (Table 4.3), where the number of men awarded the degrees has increased, and the proportions (of women) are about one-third. There is a substantial decline in the number of men and women receiving doctorates in mathematics in the last six years. Perhaps this is due to the new job opportunities for math graduates as data analysts, data scientists, etc., across the sectors. More research is required on this aspect.

The high proportions of female doctorates in STEM imply a higher value attached to women's education at the present than in the mid-twentieth century. However, with the proportions of women in research careers as miniscule, the same value is not attached to careers.

There is a continuity in the class background of the women scientists. Studies of women scientists since the 1990s show that women in science belong to middle classes and usually to a better socio-economic stratum than men (Gupta 2007b; Jaiswal 1993; Kumar 2001; Subrahmanyan 1998). This indicates a continuity in the selective access of women to higher education. Families with a higher income and educational background are more willing to send their daughters for higher education in science, and such families also appear to have a positive environment for the academic growth of girls. In a study of IITs (Gupta 2007b), about three-fourths of women and only half of men surveyed said that parents encouraged questioning rather than blind obedience. However, career aspirations from daughters are lower even among this socio-economic stratum. As many as 61% of men (and only 40% of women) reported that their parents would like them to be 'highly successful'. While expectations from the sons are in terms of professional career advancement, the daughters are expected to use their education for being 'independent', indicating the primacy of domestic roles for women. While daughters are expected to obtain good academic degrees, the lower career expectations coupled with gendered roles create hurdles for women's career aspirations.

The processes and interactions in academia construct barriers to the careers of women in science. This struggle initially unfolds during the doctoral period due to a gendered doctoral science milieu.

Gender and the informal doctoral research environment

Informal communication and interaction are integral components of the practice of science, including the doctoral process. It is a period when a doctoral student acquires training in 'tacit aspects' of research, such as learning to cope up with the vagaries of experiments (Delamont and Atkinson 2001) and to design research, write grant proposals, co-author publications, and organize people (Fox 2000). As in the West, women in India are disadvantaged in this informal learning process since women lack the same rapport as men do with their advisor, thus losing out on mentoring. A lack of interaction with colleagues and seniors leads to isolation and a dependence on merit and hard work (or the human capital). These issues are compounded by the institutional indifference to the problems of women. However, this general subordinate position of women in science assumes a different form in India, and varies in content due to the specific socio-cultural context.

Interaction with the supervisor

Given the cultural norms that limit male-female interaction, a female student is typically constrained in interacting with the male supervisor. In the study of IITs (Gupta 2007b), a majority of the women (60%) interacted formally with the supervisor for thesis work in his/her office only, while about 64% of men interacted informally with the supervisor within and outside the department. According to a woman respondent in this study (math, IIT1), 'Boys [sic] are more comfortable in interaction with the supervisor. They can approach him anytime. I have to think twice not because of the supervisor but because of my sex' (p. 520). Another respondent (math, IIT2) said, 'Not only male scholars interact more with the supervisor, the supervisor also interacts more with them. It helps in both ways' (p. 520). A lack of informal interaction with the supervisor affects the level of mentoring received from the advisor. A research scholar may have to perform extra work for the advisor, such as preparing conference presentations, writing reports, and performing experiments that have no relation to the thesis. Such work may provide a good learning experience and is important in building an informal interaction with the advisor. However, women are usually given fewer of such extra tasks, implying a lack of mentorship.

This social prejudice against women in science is also reflected in women's interaction with the supervisor. According to a woman respondent in chemistry (Gupta 2007b: 522), 'Even if the non-thesis work is given to girls [sic], boys [sic] get appreciated more. If girls [sic] do well in such tasks, guides often say, "you must have done this with the help of boys [sic]. You girls [sic] can't do this".' Some women respondents reported that faculty members were 'preferring boys [sic] in the laboratory' since they can work at odd hours and are seen to be 'more intelligent'. A woman respondent about to complete her doctorate in chemistry remarked (Gupta 2007b: 522), 'girls [sic] might get scolded much more, which puts them into depression and brings down their research level.' Male research students are preferred when work involves traveling around or if heavy work and more complex projects are involved (Subrahmanyan 2009; Maji, Mitra and Asthana 2023). Female doctoral students might be seen as a waste to the doctoral program and to the advisor, as they might leave the program to get married or they might not practice their profession after marriage (Subrahmanyan 2009).

Male scholars also have unpleasant interactions with their advisors. However, failings of women might be equated with their lack of competence and gender, affecting their self-esteem (Gupta 2007b). A dismissive and apathetic interaction of women scholars with the advisor and blatant sexism leading to a loss of self-esteem during the doctoral period have been reported by several studies (Subramanian 2017; Maji, Mitra and Asthana 2023).

These experiences are a manifestation of an unequal power relation existing between an advisor and a student coupled with gender hierarchy. Advisors are extremely powerful since they have the sole authority to decide whether sufficient quantity and quality of work has been accomplished to merit the degree (Gupta 2007b); their organizational position and networks confer them leverage over a doctoral student's future as the student requires the advisor's recommendations for career progression after obtaining the degree. This unequal power relation leads to harassment of a doctoral student at times, and in the case of a woman student, might translate into sexual harassment. The power and patriarchy are enmeshed in the persona of a male advisor. Power makes individuals less inhibited (Keltner et al. 2003). Studies on sexual harassment in Indian educational institutes are few. Subramanian (2007) found in the study of Madras University that cases of sexual harassment of women doctoral students by male advisors are common. Most women tend to brush off offensive jokes or stories, unwanted comments, queries, and suggestions (Thomas 2015).

Although common in academia worldwide, cases of sexual harassment of women doctoral students in universities were muted in India until newspaper reports started emerging at the start of the millennium (Thomas 2015). The cases of harassment from premier institutes such as the IITs/IISc have come to light only in the last few years, suggesting a greater confidence among the younger generation. Such harassment could be from a professor, student or even a canteen employee.[15] The law against sexual harassment which was enacted as late as 2013 that led to the setting up of Internal Complaints Cell (ICC) in the institutes, and the 'MeToo' movement that has emerged since 2017 across the various sectors in the country, including academia, has helped to provide visibility to the widely prevalent yet hitherto concealed issue of sexual harassment.[16] Many a times, even in the top science and technology institutes, the grievances of the victims are either ignored or covered up; improper procedures of the ICC have been reported, and the victim-blaming, especially if the perpetrator is powerful, is common. In the process of obtaining justice, the victim often faces isolation and threats, while the perpetrator receives little or no punishment (Dogra and Jayaraj 2023; Thomas 2015). Sexual harassment and the omissions of justice constitute a glaring example of absolute male power and masculinity in technical institutes.

Interaction with peers

Women doctoral students lack an informal group of friends (colleagues and seniors) for seeking help and support in academic work. They are isolated not only because they are in a minority, but also because of gendered issues in obtaining and giving help. Given the low proportion of women in most of the departments, they inevitably require the help of their male colleagues to discuss an issue or to clear up doubts that might emerge during course work or thesis. This means making appointments with them, going to their hostel, and a limited level of interaction. On the other hand, men do not require the help of their women colleagues as their interactions with their male colleagues/seniors are frequent and large in volume and scope, providing them ample access to the help of their male colleagues and seniors. Thus, it is only a one-way requirement (Gupta 2007b).

Isolation of women doctoral students impacts their work. As one respondent in mathematics said (Gupta 2007b: 523):

> I don't have much interaction with the boys [sic] because boys [sic] don't like it and they are busy amongst themselves. I feel like an outsider in the large group of boys [sic]. ... The work which can be finished in one hour, takes three days to finish because I have to search for somebody who can help.

Availability of a critical mass of women does not necessarily lead to a healthy interaction; this is either because gender-based grouping in departments occurs or a large number of women students in a department gets divided into different labs (Gupta 2007b). Competition between students of the same lab or groupings based on seniority, gender, caste, favoritism by the supervisor, regionalism, or common interests might isolate a female doctoral student further (Gupta 2007b; 2010). Epistemic cultures might also isolate women; greater interaction was noted among the theoretical groups of departments than among the experimental groups (Gupta 2010).

Lack of male empathy for problems of their female colleagues is reflected in the perception of male doctoral students that their female counterparts are 'privileged' (Gupta 2007b; 2010). Both men and women agreed in the Gupta (2007b) study, that women have fewer contacts. But men perceived women to have significantly higher access to cooperation from clerks, faculty, colleagues, and seniors, which was in complete contradiction to the perception of women (Gupta 2007b). This is partly a product of ignorance of women's problems and partly an inability of men to see their own privileged status in society. A gendered socialization since childhood is reflected in the fact that often men are unwilling to acknowledge the meritorious work of their female colleague, and that there is a belief in the intellectual superiority of men. Thus, in the same study (Gupta 2007b), while women researchers clearly rejected stereotypes about a woman's ability to pursue science, men researchers did not reject them with the same vehemence as women. About 30% of male (and only 8% female) scholars agreed that science and engineering require intellectual abilities more suited to men than women. Similarly, about one-third of the male researchers believed that men were better teachers and researchers, while only about one-tenth of the women thought so.

The seemingly 'gender-neutral' structures in the institutes are male-centric and fail to account for women. For instance, a unit of lab might be situated in a men's hostel where women students are not allowed after a certain time or a lack of security and an institute's lackadaisical response to sexual harassment (Gupta 2007b). Men get their administrative work done through the clerical staff (mostly male) more easily than women; women are less aware of the procedures and formal rules compared to men due to the absence of a large peer group (Gupta 2007b). Considering the disadvantages for women doctorates, it is not surprising that in the Gupta study (2007), while about half of the male doctoral students strongly agreed that they were confident of producing good results in their present research efforts, only one-third of women could strongly agree with it.

A positive mentoring by advisors can drastically enhance women's careers. For instance, Rosy Mondal, a cancer biologist, attributes her ability to deal with the issues of lost data and scarce samples during her degree, to her advisor at Assam University (Jayaraj 2017). She went on to obtain the prestigious DST Inspire Faculty award. Similarly, the only woman HOD of a CSIR lab at the time of a survey (Gupta 2023) was mentored by her PhD advisor. Her career at an early stage was aided by her advisor's handing her the responsibility of writing research proposals to DST and other government funding agencies, defending those proposals, writing reports, and working on high-profile assignments. This gave her the necessary experience and visibility for independent research. Incidentally, in both the cases, the advisors were male.

In pure sciences and mathematics, one usually requires postdoctoral studies for an appointment as a permanent faculty member in a reputed institute. Various institutes, including the IITs and IISc, offer postdoctoral studies. However, postdoctoral fellowships (PDFs) in India are few; for instance, at IISc, the faculty-postdoc ratio is 2.8:1 and at Stanford it is 1:1 (Naik and Megha 2018). Schemes such as that of IIT-Madras providing a PDF of one year for women with a break in career are rare.[17] There is little data on women postdoctorates and studies on the subject are few. A report based on the sample from premier institutes under eight ministries and autonomous departments under the Government of India noted that male postdoctoral fellows are 2.5 times their female peer group; whereas, at the PhD level, male enrollment is less than twice the female enrollment (SSESS 2016–2017: 30). Thus, there is probably considerable attrition of women from PhD to postdoctoral research. The postdoctoral phase might also have gendered outcomes. For instance, in biosciences, according to Bal (2005), postdoctoral studies usually take from two to ten years. This leads up to an age beyond 30 years when both men and women tend to get married. Due to the additional burden of domestic roles/childbearing, women end up having poorer professional achievements compared to men of the same age and basic qualifications. Hence, fewer women are likely to apply or are able to compete with the men.

Doctorates/postdoctorates from abroad

Although there is no formal rule, elite institutes such as the IITs have an informal practice of preferring doctorates or postdoctorates from abroad (Kurup and Raj 2022; Pushkar 2015). Many other well-established Indian universities and other higher education institutions, such as the IIITs (Indian Institutes of Information Technology), have attracted and hired a significant number of foreign PhDs (Pushkar 2015). UGC's recent rule of PhD holders from foreign universities as eligible for direct recruitment as Assistant Professors reinforces and formalizes this preference.[18] Women are less likely to travel abroad for PhDs compared to men due to family and security reasons (Campion and Shrum 2004) and have less work-experience abroad, usually in a postdoctoral research position (UNESCO 2022). The value attached to a doctorate from abroad reflects on India's science institutes. Few Indian institutes fall in the top 100 of the world rankings; good quality research occurs only in the top IITs/IISc, and they have among them a large proportion of faculty with foreign-trained PhDs. The general belief that a doctorate from India is lower in quality than one from abroad might not always be true, but the system favors foreign PhDs, which creates an automatic disadvantage for most women at the outset of their careers.

Similarly, there is a preference among students for training abroad for a *postdoctorate*, and students are strongly advised by their PhD mentors to pursue a postdoctorate overseas (Naik and Megha 2018). About 70% of students surveyed (Naik and Megha 2018) felt a need to train overseas for a faculty position in Indian academia. There is an absence of data for Indian men and women in postdoctoral programs in India and abroad. It might be assumed that it is tougher for women to move abroad for a postdoctorate, as the mobility might be constrained due to pressure to either get married or gendered-role expectations if already married. Women scientists in developing countries, including India, are subject to 'educational localism' (Campion and Shrum 2004), that is, men are more likely than women scientists to have at least some training in developed countries and to spend more time abroad for training and education. This restricts career opportunities for women. Further, women scientists in India that are trained abroad receive a negative gender shock on their return, as they have to make compromises due to gendered pressures that could affect their career advancement (Paul 2021). The salience of gender in scientific career outcomes is reported only by women and not by men scientists that are trained abroad (Paul 2021).

Thus, disadvantages accumulate for women in the doctoral phase in various ways. Lack of mentorship by the advisor adversely impacts acquisition of the necessary experience and visibility for independent research. It also affects the creation of networks with peers and professionals and inclusion in powerful networks of the advisor, thereby imposing constraints in knowing and securing good employment opportunities. Such networks, as shown in research elsewhere, establish the credibility of a candidate in hiring, promotions, awards, etc. (Cabrera and Thomas-Hunt 2007). Preference of recruiting institutes, such as doctorates from abroad, impacts women adversely as fewer women than men are able to move abroad. Even if applying with top qualifications, recruiting institutes apply a gender lens, as shown in the next chapter. While a few top institutes are making an effort to hire couples, many government institutes/labs do not prefer to do so (Gupta 2002; Jayaraj 2017) due to which either a woman compromises or the couple has a tough time in obtaining jobs, also referred to as the 'two-body problem'. In such cases a moral dilemma of choosing family over work is faced by women (Pendharkar 2016). These discriminatory patterns construct barriers for women at the beginning of their careers. The attrition of women PhDs from science is about 11% in one survey (Kurup et al. 2010). This survey also found that an inability to obtain an appropriate job was a significant factor for women who dropped out of careers after a PhD in STEM.

Academia prefers a straight career path and early achievement. However, the lack of work-life balance structures forces women to take a *break in career*. In a study (Agashe et al. 2022), 99% of the young scientists who took breaks in careers due to family reasons were women. Those who continue to do research through research projects (after doctorate) are at a disadvantage compared to those with teaching/industrial experience for a faculty position (Gangopadhyay and Das 2016). Further, after the age of 35 years, a return to academia at entry-level positions after a careerbreak is difficult, and a woman might be trapped in postdoctoral positions (Dogra and Jayaraj 2023). On the other hand, senior positions are usually filled up by internal promotions rather than through lateral hiring (Joseph and Robinson 2014). This would filter out women researchers that manage to do well (either in lesser institutes or through research projects) and apply later to the elite institutes. Thus, career trajectories off the beaten path, more common in the case of women, are discouraged. In the Kurup et al. (2010) survey of women doctorates in science, engineering, and medicine, a majority of those who were not engaged in research were doing undergraduate teaching. The GATI Report (2022) found that women often choose teaching over research because returning to a career as a researcher is difficult after marriage/children, as they are likely to lose out in terms of intellectual development and personal opportunities while balancing career with care responsibilities.

Thus, a deficit of career capital for women arises from an accumulation of disadvantages in the doctoral phase, combined with substantial gendered expectations. Also, being married to scientists does not necessarily give a boost to careers as 43% of 'women not working' (with doctorates in STEM) in the Kurup et al. (2010) study, were married to scientists engaged in research, teaching, or consultancy. Nevertheless, most women scientists married to men scientists find support from spouses (Kurup and Raj 2022; Gupta and Sharma 2002).

Gender, caste, and doctoral period

Though the proportion of the Dalit females is about 16% of the female population in the country (as per the 2011 Census), Dalit women constitute only 8.1% (AISHE Reports 2021–2022) of women enrolled in doctoral studies in all branches (AISHE Report 2019–2020; 2021–2022: Tables 11). Dalit women lag behind their male counterparts, who constitute about 10.8% of men enrolled in doctoral studies (AISHE Report 2021–2022). While the all-India ratio of female teachers per 100 males is 77, there are only 61 Dalit female teachers per 100 male Dalits (AISHE Report 2021–2022: 37). In the top engineering institutes, such as the IITs, the proportion of Dalits/Adivasis remains small despite the quota system, with only about 2.5% of the faculty in position from Dalit and 0.34% from Adivasi communities in 2018

(Ministerial Response 2018). There might be no females among the reserved category faculty in the IITs (Gupta 2023). The Dalits/Adivasis receive fewer fellowships, for instance, DST's INSPIRE faculty fellowship (supporting young talent) recipients between 2016–2020 included just 6% Dalits and less than 1% Adivasis (Paliwal 2023).

These proportions reflect the deep-seated caste prejudice in society that pollutes the lowest to the highest level of education, science institutions, and social structure. This is despite the constitutional provisions barring caste discrimination, and several state provisions to protect the Dalits/Adivasis, including reservations at student to faculty levels and national level commissions separately for the Dalits and Adivasis. Yet, there is a gap between the policy and implementation. The SC/ST cells in most IITs lack funds and space (Smart 2023). Recently IIT-Delhi has constituted an exclusive cell for monitoring the reservation policy and addressing the grievances from lower caste groups (Lakshman 2023b). However, inclusion of the OBCs or the Other Backward Class (lower than General Castes but higher than the SCs) in this obfuscates the issue (Sanghi in Lakshman 2023b). IIT Bombay's SC/ST cell recently brought out anti-casteist instructions after a suicide by a Dalit student due to toxic interaction with the fellow students and they warn of severe punishments for violations (Lakshman 2023a); although this will require effective implementation.

Caste biases were revealed in some of the statements of doctoral students in a study on an IIT (Gupta 2010: 126). For instance, a male student from chemical engineering said, 'Caste inequalities still exist. If I do well, they [peers] say it is because you are from SC'; or another student from a 'general' caste said, 'SC students are not as good as the general candidates'. Students also indicated their perception of caste biases among a few faculty members. A (non-Dalit) student of electrical engineering (EE) said, 'Lot [sic] of faculty members make sarcastic comments that nowadays SCs are privileged'; and another student from chemistry remarked, 'my guide is a Brahman and there is another faculty member in our department who is an SC. There is hostility between the two' (Gupta 2010: 126).

Protests by the Dalit/Adivasi doctoral students and the faculty from the reputed institutes illustrate their harassment. A complaint even by a Dalit faculty against harassment by professors could lead to administrative harassment and resignation by the victim (Tamilarasu 2022). Supervisors might refuse to be the guides of Dalit/Adivasi doctoral students and question their abilities (Reuters 2021). There have been several suicides by such students in these institutes (Mungekar 2023). Although usually attributed to depression, the latter is likely to be a buildup of caste-based harassment by peers or indifference of the administration. In such a scenario, Dalit/Adivasi women are at the bottom of social hierarchy and are likely to be severely disadvantaged.

Casteism plays a huge role in one's gendered experience in the doctoral phase. The Dalit/Adivasi female students are even more vulnerable to an advisor's frustrations simply because they face more severe costs for speaking out (Kondaiah et al. 2017). As Shalini Mahadev (Kondaiah et al. 2017), a doctoral student at the University of Hyderabad, notes that even though no one asked about caste, she 'stood out' in her master's class at a university because of her clothes, mannerisms, and accent; her colleagues had a better education and more comfortable life, had better knowledge, and spoke better. Further, performance expectations from the lower castes in general are low; these aspects overlap with gender since women are pressurized into being docile and are constantly judged.

Caste and patriarchy are strongly interlinked and reinforce each other (Gorringe 2018). A female researcher of a lower caste is, therefore, more vulnerable to sexual harassment than her upper caste female colleagues. Even in the progressive state of Kerala, a Dalit doctoral student faced sexual harassment, apart from the denial of basic facilities and deliberate delay by the authorities in completion of her PhD (Raghunath 2021). Generally, voices of women in science upholding gender equality in the media are from the upper castes, and silence is seen as essential by the women Dalit/Adivasi scientists for their sheer survival in the scientific community (Dogra and Jayaraj 2023). The Dalit/Adivasi women faculty face intersecting pressures of caste and masculinity within the informal environment of the educational institutes, which are discussed in the next chapter.

Conclusions

To conclude, a low number of women scientists in the elite institutes, few women in leadership positions, and lack of recognition for women scientists indicate a subordinate position of women scientists vis-à-vis the men. While research productivity is believed to be a prominent yardstick for judging merit, similar research performance might not yield similar career outcomes between men and women. This chapter shows the significance of structural and institutional factors in impacting women's careers, it shows the lacunae in our scientific and education systems and that the developing country status of India impacts women scientists. Low funding and poor quality of higher education in most universities render doctorates unemployable by the top institutes. Women faculty tend to cluster in colleges and state universities, and very few find employment in elite institutes.

The doctoral period is a period of transition from degrees to careers. However, the informal milieu for women leads to an accumulation of disadvantages. A negative evaluation by a supervisor and the belief of peers and/or supervisors that men have superior intellectual abilities, along with blatant sexism at times, serves to lower the self-esteem of female scholars.

The Indian socio-cultural norms of interaction imply low interaction and mentoring by the supervisor (usually male), affecting experience, creation of networks and contacts, resulting in a social capital deficit that could impact obtaining jobs. The prestige attached to postdoctorate/doctorate experiences abroad, and women's constrained mobility, reduce their chances for positions in the elite institutes. The doctoral period coincides with the marital and domestic roles that might create not only 'dual burden', but also age-related hurdles at the time of recruitment (women passing by the age at which institutes prefer to recruit). These disadvantages create a 'sticky floor' with women finding it hard for their careers to take off. The floor is even stickier for the Dalit and Adivasi female students due to the intersection of gender and caste.

Once recruited into careers, an important reason for women finding it difficult to rise is due to a male hegemony in numbers and position at the workplace, a lack of universalism, and notions that support a masculine culture. Beliefs about gender are reproduced in the workplace through various processes that lead to reproduction of gender in organizations. This is dealt with in the next chapter.

Notes

1 Faculty Strength Statistics, Council of IIT: https://www.iitsystem.ac.in/facultystas
2 There are 113 women fellows out of 1,150 fellows: https://fellows.ias.ac.in/
3 Ministry of External Affairs, GOI: https://indbiz.gov.in/pioneering-work-of-women-scientists-in-india-gets-a-boost/
4 Publication output: US Trends and International Comparisons, NSF. https://ncses.nsf.gov/pubs/nsb20206/publication-output-by-region-country-or-economy
5 Impact of Published Research, NSF: https://ncses.nsf.gov/pubs/nsb20206/impact-of-published-research
6 Council of IIT Council of IIT (2022), Faculty Strength Statistics. Available at: https://www.iitsystem.ac.in/facultystas; Also see, BiasWatchIndia. Estimated base rates of women faculty in institutes/universities across the sciences. Available at: https://biaswatchindia.com/base-rates-of-indian-women-faculty/
7 https://en.wikipedia.orgwikiList_of_ Shanti_Swarup_Bhatnagar_Prize_recipients.
8 Data on gender representation in INSA (2021): https://www.insaindia.res.in/pdf/ws_data.pdf
9 INAE fellowship (Indian fellows) as on Nov 1, 2021: https://www.inae.in/nomination-information/inae-fellowship/; Number of women fellows calculated from: https://www.ias.ac.in/Initiatives/Women_in_Science/Women_Fellows_of_INAE
10 BiasWatchIndia: Field-wise base rates. See https://biaswatchindia.com/base-rates-of-indian-women-faculty/
11 The Hindu Bureau: https://www.thehindu.com/news/national/parliament-proceedings-40-teaching-posts-vacant-in-iits-govt-says-in-parliament/article66255795.ece

12 Faculty shortage: Mishra 2019: https://timesofindia.indiatimes.com/education/news/why-iits-nits-are-facing-faculty-shortage-in-specialised-streams/articleshow/70359096.cms; Thakur 2022: https://timesofindia.indiatimes.com/india/several-top-iits-are-running-on-half-their-staff-strength/articleshow/96292129.cms; Agrawal 2022: https://theprint.in/india/education/budget-constraints-low-quality-candidates-why-indian-universities-are-facing-faculty-shortage/1257490/
13 See interview by Jayaraj at: https://thelifeofscience.com/2017/11/22/life-cycle-shanti-stars/; also, at https://thelifeofscience.com/2016/04/18/monica-algologist/; Also in a NIAS meeting reported in Kurup et al 2017: https://www.currentscience.ac.in/Volumes/112/10/1986.pdf
14 https://www.womenentrepreneurindia.com/viewpoint/experts-column/most-influential-indian-women-engineers-who-changed-the-game-for-women-in-technology-nwid-2373.html
15 Sexual harassment of student: http://eklavyajee06.blogspot.com/2014/06/harassment-at-edu-work-place.html; https://thewire.in/the-sciences/iisc-sexual-harassment-metoo; sexual harassment by a fellow student: https://www.ndtv.com/guwahati-news/iit-guwahati-suspends-student-arrested-in-sexual-harassment-case-2406482; harassment by professors and students against a Dalit scholar: https://www.indiatoday.in/india/story/iit-madras-professor-anticipatory-bail-sexual-assault-case-dalit-scholar-1939738-2022-04-20; by canteen worker: https://www.hindustantimes.com/india-news/canteen-worker-arrested-for-trying-to-sexually-assault-iit-madras-student-101659496207518.html#:~:text=A%2024%2Dyear%2Dold%20canteen,the%20campus%20on%20July%2024. Also see, Dogra and Jayaraj (2023) on their difficulties in collecting sexual harassment cases.
16 List of Sexual Harassers in Academia (LoSHA) was first released on October 24, 2017, on social media as a caution for students to avoid those harassers (the professors and academics), marking the advent of 'MeToo' in India. 'The List', initially included the names of about 60 highly respected academic men, 'The List' was posted on October 24, 2017 by activist Inji Pennu and an Indian student in California named Raya Sarkar, who alleged they personally confirmed every incident. A second list came out a week later that was made by women from lower caste backgrounds and included more names, bringing the total up to around 70. See https://en.wikipedia.org/wiki/Me_Too_movement_(India)
17 PDFs for women with break in career. March 03, 2017. https://www.iitm.ac.in/sites/default/files/Others/guidelines_for_post_doctoral_fellowship_for_women_with_break_in_career.pdf
18 India Today. November 29, 2018. https://www.indiatoday.in/education-today/news/story/phd-holders-from-top-foreign-varsities-are-now-eligible-for-direct-recruitment-as-assistant-professors-says-ugc-1398533-2018-11-29

References

Adler, N.J., and Harzing, A.-W. (2009). When Knowledge Wins: Transcending the Sense and Nonsense of Academic Ranking. *Academy of Management Learning and Education* 8(1): 72–95.

Agashe, D., Maheshwary, S., Pattanaik, J.K., Prakash, J., Bhatt, P., Arya, S.S., et al. (2022). Career Challenges for Young Independent Researchers in India. *Current Science* 122(2): 135–143.

AISHE Report (2021-2022). *All India Survey on Higher Education*. Government Of India, Ministry of Education, Department of Higher Education: New Delhi. Available at: https://aishe.gov.in/aishe-final-report/. Accessed on 5 September 2024.

Amarnath, V. (2011). Dilemma of Women Scientists. *Current Science* 101(3): 259.

Arora, C. (2022). Sustaining Science, Becoming Respectable. *Seminar: The Monthly Symposium* 760(December 2022: Women in Science): 26–32. Available at: https://www.india-seminar.com/semframe.html. Accessed on September 5, 2024.

Bal, Vineeta. (2005). Women Scientists in India: Nowhere near the Glass Ceiling. *Current Science* 88(6): 872–878.

Bansal, S., Viswanathan, B., and Meenakshi, J.V. (2022). Does Research Performance Explain The "Leaky Pipeline" In Indian Academia? A Study of Agricultural and Applied Economics. *Agricultural Economics* 54(2)274-288. https://doi.org/10.1111/agec.12744

Banshal, Sumit Kumar, Singh, Vivek Kumar, and Mayr, Philipp. (2019). Comparing Research Performance of Private Universities in India with IITs, Central Universities and NITs. *Current Science* 116(8): 1304–1313.

Cabrera, S.F., and Thomas-Hunt, M.C. (2007). *"Street Cred" and the Executive Woman: The Effects of Gender Differences in Social Networks on Career Advancement* [Electronic version]. Available at: https://ecommons.cornell.edu/items/eceb0818-fc03-4ba5-be45-fb1b0e4fda38/full. Accessed on September 5, 2024.

Campion, P., and W. Shrum. (2004). Gender and Science in Development: Women Scientists in Ghana, Kenya and India. *Science, Technology and Human Values* 29(4): 459–485.

Chadha, G., Krishna, S., and Tripathi, U. (2017). Blue Flower Mentoring: Interview with Vidita Vaidya. In S. Krishna and G. Chadha (eds), *Feminists and Science. Critiques and Changing Perspectives in India,* Vol. 2, Sage: New Delhi, 72–92.

Chakravartty, P., Kuo, R., Grubbs, V., and McIlwain, C. (2018). #CommunicationSoWhite. *Journal of Communication* 68(2): 254–266. https://doi.org/10.1093/joc/jqy003

Dabas, B. (2021). Publication Productivity of Women Physicists in India: A scientometrics study. *Library Philosophy and Practice* (e-journal): 5624. Available at: https://digitalcommons.unl.edu/libphilprac/5624/. Accessed on September 5, 2024.

D'Amico, Rita, Vermigli, P., and Canetto, S.S. (2011). Publication Productivity and Career Advancement by Female and Male Psychology Faculty: The Case of Italy. *Journal of Diversity in Higher Education* 4(3): 175–184. https://doi.org/10.1037/a0022570

Deccan Chronicle. (2017). Women Scientists at Defence Research and Development Organisation up 5-fold. *Deccan Chronicle.* Published Mar 9, 2017. Available at: https://www.deccanchronicle.com/nation/current-affairs/090317/women-scientists-at-defence-research-and-development-organisation-up-5-fold.html. Accessed on September 5, 2024.

Delamont, S., and Atkinson, P. (2001). Doctoring Uncertainty: Mastering Craft Knowledge. *Social Studies of Science* 31(1): 87–107.

Dogra, A., and Jayaraj, N. (2023). *Lab Hopping. A Journey to Find India's Women in Science.* Penguin: India.

DOS Report (2021–2022). *Annual Report.* Department of Space, Government of India. https://www.isro.gov.in/media_isro/pdf/AnnualReport/annual_report_english.pdf. Accessed on September 5, 2024.

Elsevier. (2016). *International Comparative Performance of India's Research Base 2009-2014.* A bibliometric Analysis. NSTMIS, Department of Science and Technology: New Delhi. April 2016. Available at: http://www.nstmis-dst.org/Pdfs/Elsevier.pdf. Accessed on September 5, 2024.

Etzkowitz, H., Kemelgor, C., and Uzzi, B. (2000). *Athena Unbound: The Advancement of Women in Science and Technology.* Cambridge University Press: New York.

European Commission (2019). *Directorate-General for Research and Innovation. She figures 2018*. Publications Office of the European Union: Luxembourg.
Fox, M.F. (1983). Publication Productivity Among Scientists: A Critical Review. *Social Studies of Science* 13(2): 285–305.
Fox, M.F. (2000). Organizational Environments and Doctoral Degrees Awarded to Women in Science and Engineering Departments. *Women Studies Quarterly* 28(1/2): 47–61.
Fox, M.F. (2004). Women in Scientific Fields: Doctoral Education and Academic Careers. Workshop on Women's Advancement, American Political Science Association. Available at: https://www.researchgate.net/publication/237578572_Women_in_Scientific_Fields_Doctoral_Education_and_Academic_Careers
Fox, M.F. (2006). Women and Academic Science: Gender, Status, and Careers. In C. Marzabadi, V. Kuck, S. Nolan and J. Buckner (eds), *Dissolving Disparity, Catalyzing Change: Are Women Achieving Equity in Chemistry?*, Oxford University Press: New York, 17–27.
Garg, K.C., and Kumar, Suresh. (2014). Scientometric Profile of Indian Scientific Output in Life Sciences with a Focus on the Contributions of Women Scientists. *Scientometrics* 98(3): 1771–1783.
Gangopadhyay, R., and Das, B. (2016). DST WOS-A: The Scenario from Recipient's Perspective. *Current Science* 111(8): 1307–1309.
GATI Report (2022). *Gender Advancement for Transforming Institutions. A Baseline Study*. February 2022. British Council and DST. Available at: https://gati.dst.gov.in/resources/wspublic;jsessionid=CE0B091C8F66AAC5B2C520A68D567495/resource-material/GATI-Report-February_2023.pdf. Accessed on September 5, 2024.
Godbole, R., and Ramaswamy, R, eds. (2008). *Lilavati's Daughters: The Women Scientists of India*. Indian Academy of Sciences: Bangalore, India.
Godbole, R.M., and Ramaswamy, R. (2015). Women Scientists in India. In R.M. Godbole and R. Ramaswamy (eds), *Women in Science and Technology in Asia*. Indian Academy of Sciences: Bangalore, 67–84. Available at: https://www.ias.ac.in/public/Resources/Initiatives/Women_in_Science/AASSA_India.pdf. Accessed on September 5, 2024.
Gonslaves, Timothy A. (2023). Is BTech A Good Career Option for Girls? Former IIT Mandi Director explains. *The Indian Express*, February 11, 2023. Available at: https://indianexpress.com/article/education/a-lesson-from-iit-is-btech-good-girls-timothy-gonsalves-jee-main-advanced-8419115/. Accessed on September 5, 2024.
Gorringe, H. (2018). Afterword, Gendering Caste, Honor, Patriarchy and Violence. *South Asia Multidisciplinary Academic Journal* [online]. Available at: https://journals.openedition.org/samaj/4685. Accessed on September 5, 2024.
Gupta, N. (2007a). How Critical is Critical Mass? Implications of the Minority Status of Women in Science in India. *Current Science* 93(6): 766–768.
Gupta, N. (2007b). Indian Women in Doctoral Education in Science and Engineering, A Study of Informal Milieu at The Reputed Indian Institutes of Technology. *Science Technology and Human Values* 32(5): 507–533.
Gupta, Namrata. (2010). Doctoral Research Environment in an Indian Institute of Higher Learning in Science and Technology. *Science Technology and Society* 15(1): 113–133.
Gupta, Namrata. (2016). Perceptions of the Work Environment: The Issue of Gender in Indian Scientific Research Institutes. *Indian Journal of Gender Studies* 23(3): 437–466.
Gupta, Namrata. (2017). Gender Inequality in The Work Environment: A Study of Private Research Organizations in India. *Equality, Diversity and Inclusion: An International Journal* 36(3): 255–276.

Gupta, N. (2023). Women in STEM in India: Understanding Challenges through Social Constructionist Perspective. *American Behavioural Scientist* 67(9): 1084–1103.
Gupta, B.M., Kumar, Suresh, and Aggarwal, B.S. (1999). A Comparison of Productivity of Male and Female Scientists of CSIR. *Scientometrics* 45(2): 269–289.
Gupta, N., and Sharma, A.K. (2002). Women Academic Scientists in India. *Social Studies of Science* 32 (5–6): 901–915.
Gupta, N., and Sharma, A.K. (2003). Gender Inequality in The Work Environment at Institutes of Higher Learning in Science and Technology in India. *Work, Employment and Society* 17(4): 597–616.
Hasan, S.A., Sharma, Mahender Kumar, Khilnani, Sushila, and Luthra, R. (2012). Research Productivity of Female Research Scholars and Their Migration Pattern in Pursuit of Higher Education and Research. *Current Science* 103(6): 611.
Hofstra, B., Kulkarni, Vivek V., Galveza, Sebastian Munoz-Najar, Heb, Bryan, Jurafsky, Dan, and McFarland, Daniel A. (2020). The Diversity–Innovation Paradox in Science. *PNAS* 117(17): 9284–9291. Available at: https://www.pnas.org/doi/full/10.1073/pnas.1915378117. Accessed on September 5, 2024.
IAP (2016). *Women in S&T: A Vision Document.* Inter-Academy Panel on 'Women in Science' in India. NASI: Allahabad. Available at: http://insaindia.res.in/pdf/ws_anx_1c.pdf. Accessed on September 5, 2024.
Jaiswal, R.P. (1993). *Professional Status of Women: A Comparative Study of Women in Science and Technology.* Rawat Publications: Jaipur.
Jayaraj, N. (2016). *Smitha Hegde and Her Indomitable Ferns.* The Life of Science. July 11, 2016. Available at: https://thelifeofscience.com/2016/07/11/smitha-hegde-and-her-indomitable-ferns/. Accessed on September 5, 2024.
Jayaraj, N. (2017). *Rosy's Red Alert to the North-East.* The Life of Science. Posted on November 29, 2017. Available at: https://thelifeofscience.com/2017/11/29/rosys-red-alert-north-east/. Accessed on September 5, 2024.
Jolly, P. (2022). Catalysing Change for Gender Advancement in Science. *Seminar: The Monthly Symposium* 760(December 2022: Women in Science): 33–38. Available at: https://www.india-seminar.com/semframe.html. Accessed on September 5, 2024.
Joseph, M., and Robinson, A. (2014). Policy: Free Indian Science. *Nature* 508(7494): 36–38. Available at: https://www.nature.com/news/policy-free-indian-science-1.14956#/b2. Accessed on September 5, 2024.
Kanter, R. (1977). *Men and Women of the Corporation.* Basic Books: New York.
Keller, E.F. (2001). The Anomaly of a Woman in Physics. In Mary Wyer, M. Barercheck, D. geisman, H.O. Ozturk and M. Wayne (eds), *Women, Science, And Technology: A Reader in Feminist Science Studies,* Routledge: New York, 23–30.
Keltner, Dacher, Gruenfeld, Deborah H., and Anderson Cameron. (2003). Power, Approach, and Inhibition. *Psychological Review* 110(2): 265–284.
Knorr-Cetina, K.D. (1981). *The Manufacture of Knowledge. An Essay on the Constructivist and Contextual Nature of Science.* Pergamon Press: Oxford.
Kondaiah, Bittu Karthik, Mahadev, Shalini, Grace, Maranatha, and Wahlang, Tham. (2017). The Production of Science Bearing Gender, Caste and More. *Economic and Political Weekly* 52(17): 73–79.
Kong, H., Martin-Gutierrez, S., and Karimi, F. (2022). Influence of the First-Mover Advantage on The Gender Disparities in Physics Citations. *Communications Physics* 5, Article no. 243): 1-11. Available at: https://www.nature.com/articles/s42005-022-00997-x. Accessed on September 5, 2024.

Kozlowski, D., Larivière, Vincent, Sugimoto, Cassidy R., and Monroe-White, Thema. (2022). Intersectional Inequalities in Science. *PNAS* 119(2): e2113067119. Available at: https://www.pnas.org/doi/10.1073/pnas.2113067119. Accessed on September 5, 2024.

Krishna, V.V. (2014). A Disappointment for Scientific Community. *The Hindu,* July 26, 2014. Available at: https://www.thehindu.com/opinion/op-ed/a-disappointment-for-scientific-community/article6250153.ece. Accessed on September 5, 2024.

Krishna, V.V., and Chandra, N. (2009). *Knowledge Production and Knowledge Transfer: A Study of Two Indian Institutes of Technology (IIT, Madras and IIT, Bombay).* ARI Working Paper Series, Asia Research Institute, Working paper 121.

Kumar, Neelam. (2001). Gender Stratification in Science: An Empirical Study in the Indian Setting. *Indian Journal of Gender Studies* 8(1): 51–67.

Kurup, A., Maithreyi, R., Kantharaju, B., and Godbole, R. (2010). Trained Scientific Women Power: How Much Are We Losing and Why? *IAS–NIAS Research Report.* Available at: http://eprints.nias.res.in/142/1/IAS-NIAS-Report.pdf. Accessed on September 5, 2024.

Kurup, Anitha, Mathew, L., and Singh, T. (2017). Women in STEM Disciplines. Meeting Report. *Current Science* 112(10): 1986–1987.

Kurup, A., and Raj, A. (2022). Mapping the Life Trajectories of Women Scientists in India: Successes and Struggles. *Current Science* 122(2): 144–148.

Lakshman, A. (2023a). IIT-Bombay Sends Missive on Anti-Discrimination Guidelines Ahead of Arrival of New Batch. *The Hindu*, July 31, 2023. Available at: https://www.thehindu.com/news/cities/mumbai/iit-bombay-sends-missive-on-anti-discrimination-guidelines-ahead-of-arrival-of-new-batch/article67142466.ece#:~:text=These%20include%20disallowing%20students%20from,prohibits%20the%20spreading%20of%20casteist%2C. Accessed on September 5, 2024.

Lakshman, A. (2023b). IIT-Delhi Constitutes SC/ST-OBC Cell, Notifies Mandate, Terms of Reference. *The Hindu*, March 16, 2023. https://www.thehindu.com/news/national/iit-delhi-constitutes-scst-obc-cell-notifies-mandate-terms-of-reference/article66627435.ece. Accessed on September 5, 2024.

Laudel, G. (2006). The "Quality Myth": Promoting and Hindering Conditions for Acquiring Research Funds. *Higher Education* 52(3): 375–403.

Lee, S., and Bozeman, B. (2005). The Impact of Research Collaboration On Scientific Productivity. *Social Studies of Science* 35(5): 673–702.

Lemoine, W. (1992). The Frequency Distribution of Research Papers and Patents According to Sex. The case of CSIR, India. *Scientometrics* 24(2): 449–469.

Lincoln, A.E., Pincus, S., Koster, J.B., and Leboy, P.S. (2012). The Matilda Effect in Science: Awards and Prizes in the US, 1990s and 2000s. *Social Studies of Science* 42(2):307–320.

Liu, F., Rahwan, T., and AlShebli, B. (2023). Non-White Scientists Appear on Fewer Editorial Boards, Spend More Time Under Review, and Receive Fewer Citations. *PNAS* 120(13): e2215324120. https://doi.org/10.1073/pnas.2215324120. Accessed on September 5, 2024.

Long, Scott J., and Fox, Mary F. (1995). Scientific Careers: Universalism and Particularism. *Annual Review of Sociology* 21: 45–71.

Maji, S., Mitra, S., and Asthana, M.K. (2023). 'Treading the No Woman's Land': The Gender-STEM Dynamics in Higher Education in Premier Institutions of India. *European Journal of Engineering Education* 48(3):422–447.

Marginson, S., and Xu, Xin. (2021). *Moving Beyond Centre-Periphery Science: Towards an Ecology of Knowledge.* Working paper no. 63. April 2021. Centre for Global Higher Education working paper series, April 2021.

Merton, R.K. (1942). The Normative Structure of Science. In N.W. Storer (ed), *The Sociology of Science,* University of Chicago Press: Chicago, 267–278.

Ministerial Response (2018). Lok sabha starred question No. 267. To Be answered on 31.12.2018. Indian Institutes Of Technology. Available at: https://sansad.in/getFile/loksabhaquestions/annex/16/AS267.pdf?source=pqals. Last accessed on November 15, 2024.

Mungekar, B. (2023). Data on Caste-Related Deaths at Educational Institutes Is Telling. *The Hindustan Times*, May 9, 2023. Available at: https://www.hindustantimes.com/cities/mumbai-news/data-on-caste-related-deaths-at-educational-institutes-is-telling-101678303267238.html. Accessed on September 5, 2024.

Naik, S., and Megha, K. (2018). The Curious Case of The Missing Indian Postdoc. Posted on September 7, 2018 in Career Development, Policy and Postdoc. *IndiaBioscence*. Available at: https://indiabioscience.org/columns/opinion/the-curious-case-of-the-missing-indian-postdoc. Accessed on September 5, 2024.

Nature (2020). Accounting for Sex and Gender Makes Science Better. *Nature*, Editorial, 588(10 December 2020): 196. Available at: https://www.nature.com/articles/d41586-020-03459-y. Accessed on September 5, 2024.

NIRF (2018). National Institutional Ranking Framework. MHRD. India rankings 2018. Available at: https://www.nirfindia.org/Rankings/2018/Ranking2018.html. Accessed on September 5, 2024.

Palackal, A., M. Anderson, and B. Paige Miller. (2007). Internet Equaliser? Gender Stratification and Normative Circumvention in Science. *Indian Journal of Gender Studies* 14 (2): 231–257

Paliwal, A. (2023). How India's Caste System Limits Diversity in Science — in Six Charts. *Nature* 613(January 12, 2023): 230–234.

Paswan, J., and Singh, V.K. (2020). Gender and Research Publishing Analyzed Through the Lenses of Discipline, Institution Types, Impact and International Collaboration: A Case Study from India. *Scientometrics* 123(1): 497–515.

Paul, A. (2021). Higher Education in North East India: Trends, Opportunities and Challenges. In Higher Education in North East India, edited by Dr. S. Elika Assumi, Testo College, Assam, 1-14. Available at: https://neissr.ac.in/wp-content/uploads/2024/02/fr.c.panto_.pdf. Last accessed on November 14, 2024.

Paul, Anju Mary. (2021). *Asian Scientists on the Move: Changing Science in a Changing Asia.* Cambridge University Press: Cambridge.

Pendharkar, V. (2016). *Women in Science, Engineering and Research' Conclave 2016: A report.* IndiaBioscience. Online. April 26. 2016. Available at: https://indiabioscience.org/columns/indian-scenario/women-in-science-engineering-and-research-conclave-2016-a-report. Accessed on September 5, 2024.

Pradeep, T. (2018). How Far Does a PhD Go? *The Hindu*. September 20, 2018. Available at: https://www.thehindu.com/opinion/op-ed/how-far-does-a-phd-go/article24988294.ece. Accessed on September 5, 2024.

Prathap, G., and Gupta, B.M. (2009). Ranking of Indian Engineering and Technological Institutes for Their Research Performance during 1999–2008. *Current Science* 97(3): 304–306.

Pushkar (2015). Should India's Elite Universities Be Partial to Foreign PhDs in Faculty Hires? *The Wire*. Education. December 13, 2015. Available at: https://thewire.in/education/should-indias-elite-universities-be-partial-to-foreign-phds-in-faculty-hires. Accessed on September 5, 2024.

Raghunath, A. (2021). Kerala Dalit Student's Fight Against Discrimination Towards Her Finds Support. *Deccan Herald*, November 3, 2021. Available at: https://www.deccanherald.com/national/south/kerala-dalit-students-fight-against-discrimination-towards-her-finds-support-1047094.html. Accessed on September 5, 2024.

R&D Statistics (2000–2001). *Research and Development Statistics.* Ministry of Science and Technology, New Delhi. May 2002. Available at: http://14.139.60.153/bitstream/123456789/7623/1/Research%20and%20Development%20Statistics%202000-01%20Department%20of%20science%20and%20%20technology%20D11527.pdf. Accessed on September 5, 2024.

R&D Statistics at a Glance (2022–2023). *Research and Development Statistics.* Ministry of Science and Technology, New Delhi. March 2023. Available at: https://dst.gov.in/sites/default/files/R%26D%20Statistics%20at%20a%20Glance%2C%202022-23.pdf. Accessed on September 5, 2024.

Reuters (2021). PhD Student's Fight Against Caste Discrimination Raises Dalit Hopes. *Deccan Herald*, Reuters, Chennai, November 30, 2021. Available at: https://www.deccanherald.com/national/south/phd-students-fight-against-caste-discrimination-raises-dalit-hopes-1056028.html. Accessed on September 5, 2024.

Saha, S., and Ray, A.S. (2015). *Science Research and Knowledge Creation in Indian Universities: Theoretical Perspectives and Econometric Evidence.* CITD Discussion Paper No. 15–10, Centre for International Trade and Development, Jawaharlal Nehru University, New Delhi, August 2015. Available at: https://www.jnu.ac.in/sites/default/files/citd/DP10_2015.pdf. Accessed on September 5, 2024.

Schiebinger, L., Klinge, I., Sánchez de Madariaga, I., Paik, H.Y., Schraudner, M., and Stefanick, M., eds. (2011–2021). *Gendered Innovations in Science, Health & Medicine, Engineering and Environment.* Available at: http://genderedinnovatio.s.stanford.edu/index.html. Accessed on September 5, 2024.

Shah, C. (2017). Body, Reproduction and Technology: Local subversions and Global Regressions. In S. Krishna and G. Chadha (eds), *Feminists and Science. Critiques and Changing Perspectives in India,* Vol. 2, Sage: New Delhi, 121–146.

Sharma, K. (2018). These Universities Are Less Than a Decade Old but Are Facing a Severe Shortage of Teachers. *The Print*, 29 May, 2018. Available at: https://theprint.in/governance/these-universities-are-less-than-a-decade-old-but-are-facing-a-severe-shortage-of-teachers/63729/. Accessed on September 5, 2024.

Shiva, V. (2016). *Staying Alive: Women, Ecology, and Development.* North Atlantic Books: London, Reprint edition (1 March 2016).

Smart, P. (2023). In Most IITs Across India, SC/ST Cells Barely Have Any Activity or Funds. *Indian Express*, May 4, 2023. Available at: https://indianexpress.com/article/education/in-most-iits-across-india-sc-st-cells-barely-have-any-activity-or-funds-8588369/. Accessed on September 5, 2024.

Somasekhar, M. (2018). From Manning Missiles To HR, Women Power Rising in Defence R&D, Production. *Businessline*, March 7, 2018. Available at: https://www.thehindubusinessline.com/news/variety/from-manning-missiles-to-hr-the-rising-women-power-rising-in-defence-rd-production/article22970516.ece. Accessed on September 5, 2024.

SSESS (2016–2017): *Final Report Status of Women in Science among Select Institutions in India: Policy Implications.* SSESS: Kolkata. Available at: https://drive.google.com/file/d/1wTmKM--4rVR-P6xaipcD2OD52PJ2a1MX/view. Accessed on September 5, 2024.

S&T Indicators (2023). *S&T Indicators Tables 2022–23.* March 2023. Department of Science & Technology. Government of India: New Delhi. Available at: https://dst.gov.in/document/reports/st-indicators-tables-2022-23. Accessed on September 5, 2024.

Subrahmanyan, L. (1998). *Women Scientists in the Third World: The Indian Experience.* Sage: New Delhi.

Subrahmanyan, Lalita. (2009). Women in Science in India. Has Feminism Passed them By? In Neelam Kumar (ed), *Women and Science in India,* OUP: New Delhi, 178–205.

Subramanian, J. (2007). Perceiving and Producing Merit, Gender and Doing Science in India. *Indian Journal of Gender Studies* 14(2): 259–284.

Subramanian, J. (2015). Unravelling the 'Gender-Merit' Conundrum: Do Women Deserve to do Science in India. In S. Krishna and G. Chadha (eds), *Feminists and Science. Critiques and Changing Perspectives in India,* Vol. 1, STREE: Kolkata, 22–51.

Subramanian, J. (2017). Mathematics to Mathematics Education: A Telling Trajectory. *Economic and Political Weekly* 52(17): 52–60.

Sur, A. (2009). Dispersed Radiance. Women Scientists in C.V. Raman's Laboratory. In Neelam Kumar (ed), *Women and Science in India,* OUP: New Delhi, 98–136.

Swarup, A., and Dey, T. (2019). Women in Science and Technology. An Indian Scenario. *Current Science* 119(5): 744–748. https://doi.org/10.18520/cs/v119/i5/744-748

Tamilarasu, P. (2022). TN Is Progressive, But IIT Madras Is Casteist: Former Teacher Vipin Veetil. *The Federal,* February 9, 2022. Available at: https://thefederal.com/states/south/tamil-nadu/tn-is-progressive-but-iit-madras-is-casteist-former-teacher-vipin-veetil/

Teich, E.G., Kim, Jason Z., Lynn, Christopher W., et al. (2022). Citation Inequity and Gendered Citation Practices in Contemporary Physics. *Nature Physics* 18(October 2022): 1161–1170.

Thakkar, P. (2008). Anandi Gopal. Anandibai Joshi. In R. Godbole and R. Ramaswamy (eds), *Lilavati's Daughters: The Women Scientists of India,* Indian Academy of Sciences: Bangalore, 13–16.

Thelwall, Mike. (2018). Do Females Create Higher Impact Research? Scopus Citations and Mendeley Readers for Articles from Five Countries. *Journal of Informetrics* 12(4): 1031–1041.

Thelwall, Mike, Bailey, Carol, Makita, Meiko, and Madalli, Devika P. (2019). Gender and Research Publishing in India: Uniformly High Inequality? *Journal of Informetrics* 13(1):118–131.

Thomas, Anju. (2015). Incidents of Sexual Harassment at Educational Institutions in India: Preventive Measures and Grievance Handling. *International Journal of Recent Advances in Multidisciplinary Research* 2(3): 317–322.

UIS (2019). *UNESCO Institute for Statistics.* Fact Sheet no. 55. June 2019. Available at: https://uis.unesco.org/sites/default/files/documents/fs55-women-in-science-2019-en.pdf. Accessed on September 5, 2024.

UNESCO (2022). *A Braided River: The Universe of Indian Women in Science.* By C. Coley, A. Dhillon, C. Gressel. UNESCO: New Delhi.

Upadhye, Rekha P., Girap, P.V., Tewari, S., Ashok, T., Bhanumurthy, K., and Sinha, R.K. (2014). Journal Publication Productivity of Women Scientists at Bhabha Atomic Research Centre, India. *International Journal of Nuclear Knowledge Management* 6(3): 268–281.

Vaid, M. (2019). *ISRO's Magnificent Women and Their Flying Machines.* Speaking Tiger Books LLP: New Delhi.

World Bank. (2024a). *Research and Development Expenditure (% of GDP).* The World Bank Group. Available at: https://data.worldbank.org/indicator/GB.XPD.RSDV.GD.ZS. Accessed on September 5, 2024.

World Bank. (2024b). *Researchers in R&D (per Million People). India.* The World Bank Group. Available at: https://data.worldbank.org/indicator/SP.POP.SCIE.RD.P6?locations=IN. Accessed on September 5, 2024.

Zainab, A.N. (1999). Personal, Academic and Departmental Correlates of Research Productivity: A Review of Literature. *Malaysian Journal of Library & Information Science* 4(2): 73–110.

5

GENDER AND STEM RESEARCH

Part 2

Introduction

Few women scientists obtain positions of leadership or awards, as discussed in the previous chapter. However, studies on well-known institutes (e.g., Gupta and Sharma 2002; Gupta 2016; UNESCO 2022) show that there is no overt discrimination against women in the formal environment, that is, the rules of recruitment and promotions do not discriminate against women. Nevertheless, there are barriers in the informal environment that construct a glass ceiling. The barriers to obtaining recognition at the higher levels of hierarchy among women scientists are partly a product of biases in international and national systems of science and a flawed notion of 'merit' as discussed in the earlier chapter. Partly, these barriers are constructed in everyday practices at scientific institutes and labs so that accumulation of 'merit' is difficult for women scientists.

This chapter focuses on how gender is reproduced and experienced in science and technology organizations. Scientific research is conducted in organizations and the latter are believed to be rational and gender-neutral. However, the belief in the rationality of organizations masks a gender culture that hinders the full utilization of the potential of women scientists. Although there exists a male dominance in scientific research organizations, this dominance is not a simple function of a lack of critical mass of women, but a product of patriarchal relations in society woven with the context of the practice of science and the organizational milieu. This chapter first focuses on how *male privilege is constructed* through the structural and normative factors that favor men in science; and then analyzes the *construction of female disadvantage* due to certain socio-cultural norms that devalue

DOI: 10.4324/9781003596660-6

women in public spaces, leading to a gendered position of women in science. The structural and normative factors that construct male privilege differ from profession to profession in details. Therefore, issues that discriminate against women scientists are different compared to those in other professions. However, the second part, construction of female disadvantage, is likely to be common to most professions that are both male dominated and 'masculine'. These norms, behaviors, and practices lead to marginalization of women at the workplace. There is a common 'triple burden' on women in science across cultures (Gupta et al. 2004). However, the norms, behaviors, and practices constructing a 'triple burden' reflect a national institutional system of science and national cultural norms.

Relationship between organizational and the Indian socio-cultural context

Patriarchy and hierarchy are significant aspects of the Indian society. They contribute to a masculine environment at the workplace. For instance, the patriarchal ideology of separation of spheres, with a woman as the fulcrum of the domestic sphere, continues to exist in the minds of employers and male colleagues at work (Patel and Parmentier 2005). Organizations typically assume that a working woman is placing family before work and the 'dual burden' is a woman's problem. As a result, even in multilateral organizations and universities, which are liberal spaces, dual income couples face the daunting task of childcare and elderly care (Grover et al. 2015). There is a prevalence of traditional gender role ideologies, and people do not think that women need support in managing the dual responsibilities. This ideology extends to public spaces and women find scarce organizational support for family responsibilities (Ramadoss and Rajadhyaksha 2012).

'Hierarchy' has been a prominent feature of the traditional Indian social structure dominated by caste hierarchy. Although public spaces are open to all, India exhibits a high 'power distance', a cultural dimension noted by Hofstede (2001), scoring high on this dimension, 77 (compared to the world average of 56.5). Power distance is the degree to which members of an organization or society expect and agree that power should be shared unequally. A high power distance indicates an appreciation for hierarchy and a hierarchical structure in society and organizations. Studies of hierarchical patterns in Indian industrial firms indicate that Indian business culture is hierarchical, with boss-subordinate relations that are highly valued (Sinha and Kanungo 1997; Heuer 2006). India's red-tape is an 'epidemic'; its bureaucratic morass impedes research and innovation (Nature 2015). Research has shown that women in science benefit from formalized structures specifying duties, expectations, and criteria of performance evaluation and disseminating information evenly (Long and Fox 1995) and that hierarchy and bureaucratic structure by themselves are not detrimental to gender equity (Roth and Sonnert 2011). However, as shown in a study (Gupta 2016) and in this chapter, inefficiencies of procedures, a hierarchical culture that values contacts with those in power, combined with Indian norms of interaction, affect women scientists adversely and help to perpetuate the male dominance in the system.

Since the Nehruvian days, the 'power elite' in science or the scientists close to political establishment, (for instance, the Prime Minister's advisory council or secretary-DST) was drawn or constituted from the MOSA (Mission Oriented Science Agencies, such as the CSIR, the Department of Atomic Energy, the Defence Research and Development Organisation, etc.), rather than from the academic settings or the universities (Krishna 2001: 244). However, it is often alleged that 'most scientists at the national laboratories merely consider themselves as just any other government employee with a desire to wield bureaucratic power' (Pandey 2007: 7). In recent times, the power elite are also being drawn from academic institutes, such as IISc and the older IITs, that are autonomous. The bureaucratic nature of most science institutes in post-colonial India and the fact that traditionally Indian society has been a caste society and has a hierarchical culture, renders the system, composed of both scientific and non-scientific members, excessively respectful of those in power with members seeking contacts with the powerful. Indian scientists (Joseph and Robinson 2014) and scholars (such as Kumar 2009; Subramanian 2007; Arora 2022) agree that science and its institutions exhibit feudal and authoritarian values prevalent in the Indian society.

It is in this background that the relationship between gender and scientific organizations is discussed. This relationship takes two forms: a, construction of male privilege; and b, construction of female disadvantage. Although separated analytically here, the two aspects are interlinked.

I. Construction of male privilege

A masculine culture in science is a product of a lower proportion of women coupled with an unequal power relation, evident in the hegemony of men in positions of power and dominance of men's practices. Dresselhaus, a Physicist, believed that women will experience lower career obstacles once their numbers reach 10–15%. However, this proportion is no longer seen as effective, and parity is the target (Etzkowitz et al. 2020). Further, in an organization, the consequences of 'tokenism' or minority status (Kanter 1977), such as performance pressure to overachieve (heightened visibility), barriers in informal socialization (boundary heightening), and casting women in stereotyped roles (role encapsulation), are not merely due to low proportions but also a low social position of a minority. This is evident in the studies that find men in female-dominated professions experiencing enhanced career opportunities or a 'glass escalator' (Williams 1992).

A male nurse might be presumed to be a physician or orderly (upgraded or downgraded), but a female academic administrator is often presumed to be a personal assistant (downgraded) (Etzkowitz 1971). It is not merely low proportions of women but also gender hierarchy in society and in science that create barriers for women scientists.

As seen earlier, women scientists in India constitute about 8–20% in many reputed government sector research institutes and organizations. However, even in professional fields such as biological sciences where percentage of women is higher, men occupy most positions of power. A minority status of women overlaps with the norms and practices that construct privileges and power for men in science organizations. These norms and practices include:

a. Gendered socialization and networking.
b. Informal practices at the workplace.
c. The concept of an ideal worker/scientist.
d. Regarding women as 'privileged'.

The section also looks at two other aspects related to gender, minority status and science organizations: the working of caste and masculinity, and construction of masculinity in the private research labs.

Gender and informal interaction

Mentoring, networking, and having contacts are regarded as significant in the professional development in science academia (Etzkowitz et al. 2000; Levinson 1978; Reskin 1978). This is because 'who you know' web of contacts and relationships create 'social capital'. This capital is a vital source of information on sources of materials, data, or services. Networking within and outside one's workplace helps to acquire tacit knowledge on how to get ahead, for instance, tips about funding/award opportunities, potential employees, or job prospects, and current research of other scientists. Further, it makes possible an exchange of ideas, resources, and collaborations. A scientist, through networks outside the workplace, introduces his/her doctoral/postdoctoral students to each other or to other scientists for future employment or research. Social capital establishes credibility of a candidate in hiring, promotions, awards, etc. (Cabrera and Thomas-Hunt 2007). Wennerås and Wold (1997) found that the distribution of fellowships was made to persons known by the committee members (at least one member).

Research elsewhere shows that women lack networking both within and outside their workplace. This is due to two reasons: First, women are less incorporated into informal chat groups, tea clubs, and old boys' networks due to a 'boundary heightening' which leads to exclusion of the minority in socialization (Kanter 1977). Second, in forming network ties, individuals tend to interact with those who have similar attributes such as race, gender, and religion (McPherson et al. 2001). So, men tend to network with men, also referred to as gender homophily. Belle, Smith-Doerr and Brien (2014), in research on networks of STEM faculty, found gender homophily, which was greater with a lack of critical mass, and women in departments *with critical mass* (15% or more) reported greater satisfaction with collaboration opportunities. A lack of networking creates a deficit of information, collaboration, and mentors and an excessive dependence on human capital (Etzkowitz et al. 2000). Through exclusion from powerful male networks, women are relegated to the periphery of scientific activity, which enhances a masculine character of the institutions.

In India, gender homophily and tokenism are enhanced due to interaction norms (upper-caste patrifocal restrictions on male-female interaction), and a hierarchical culture (Gupta and Sharma 2003; 2002; Gupta 2016). According to a woman scientist (E2, CSIR2), 'Informal interaction is there but mostly with women colleagues ...' A female scientist (Scientist E2, CSIR1) remarked, 'You can't be pally with men scientists ...' (Gupta 2016: 451). Women also interact less due to a lack of time and wanting to leave after office hours have ended. However, as Gautam (male, scientist, CSIR1) said, 'In the Indian scenario, more than professional contacts, informal contacts are likely to help you. I interact informally in the Institute mainly at tea' (Gupta 2016: 451). Men scientists also tend to stay beyond office hours. Women usually do not form part of the informal meetings after office hours when names for various committee members are decided (Gupta and Sharma 2002). Jayaraj (2017) found that where the boss is a male, and women are excluded from after-work hangouts, smoking breaks, outstation field trips, etc. a male usually had a higher chance for appropriating most opportunities, whether promotion or attending an international conference. While e-communications help to circumvent patrifocal restrictions and facilitate contacts abroad, women have fewer contacts within their institute (Gupta and Sharma 2003) and even within their departments (Palackal et al. 2007) compared to men scientists.

Informal contacts within the institutes lead to interpersonal relations that often take the form of 'pull', 'prejudice', and 'favoritism'. An overwhelming proportion (more than 70%) of women scientists admitted in a study that there were powerful groups in their department that influenced decisions, and that they (these women) were not part of any group (Gupta and Sharma 2003).

Networking within one's institute helps in creating opportunities for mentorship. This is more important for the women scientists at the junior level who need help in understanding the system and the opportunities available for research and information on how to move ahead in one's career. However, women have few mentors, and this is acknowledged by male scientists also (Gupta 2016). Further, having mentors and networking with seniors helps junior scientists to counter hierarchical culture. It creates a rapport with seniors or the influential, which provides a faster movement ahead in the institute hierarchy. A lack of such rapport with the seniors might lead to delays in promotions. A scientist in a study (Gupta 2016: 450) remarked, 'It is easier if you have a godfather; otherwise, you might be late (in getting promotion) by three to five years' (male, Scientist F in CSIR2). Without contacts with seniors, a junior scientist is vulnerable to the politics of those in higher positions. This is more so in the case of women due to an overlapping of gender and the lack of social capital.

For example, in CSIR2, Poonam (female, Scientist F), who said, 'I am not a member of any committee and have never been; I was always isolated ...', was caught between two HODs when she was to be promoted from E2 to F-level scientist. 'It took several years because one HOD left and the new one wanted to get his own back on the first one; I was the soft target. I could not trouble strong people, I don't fight back' (Gupta 2016: 450).

Contacts and networks with other scientists *outside the workplace* generate social capital, and their deficit affects career opportunities. Almost all the women scientists interviewed in a study (Gupta and Sharma 2003) reported that merit alone is insufficient for success in a scientific career; that geographical mobility, and contacts and networking with other scientists are important. Interactional norms, family constraints, and structural constraints in traveling due to security reasons constrain women in networking through participation in professional gatherings outside the workplace, even within their country (Gupta and Sharma 2003). The idea of childcare (or childcare grants) at workshops/meetings in India is nascent and is yet to take off. Further, women scientists often find themselves excluded from such gatherings. BiasWatchIndia, an on-line initiative founded in 2020, which documents women's representation, particularly in offline or online meetings, conferences, and talks, regularly reports on glaring underrepresentation. Dr Vaishnavi Ananthanarayanan, one of the founders, noted that there is an 'implicit and explicit discrimination against having women speakers. Typically, men would make it an "old boys club" and invite their friend' (Amin 2021). A lesser networking of women with men who form the bulk of the scientific community leads to lesser visibility and lesser recognition than men, which in turn implies lesser contacts and networking, and so the vicious circle continues (Gupta and Sharma 2003), thus strengthening the glass ceiling.

Age and seniority have positive consequences for the women scientists' interaction and networking. Women scientists that are aged 50 years or more are more mobile as they reported higher participation in seminars and conferences (due to higher research activity, fewer family responsibilities) than their younger counterparts (Gupta and Sharma 2003). Further, a higher age and rank lowers the socio-normative emphasis on sex segregation, providing greater leeway for mixed interaction. As Saraswati (female, Scientist G, CSIR2), who was 57 years old, remarked, 'Now I have become senior, I don't mind barging into a group of drinking people [male colleagues], but it was difficult until a few years ago' (Gupta 2016: 451). However, whether such late networking helps in catching up with career lags encountered earlier is doubtful, and only a few older women scientists reach leadership positions or obtain recognition.

Gendered practices in formal work

No government institute makes a distinction between men and women regarding the allotment of various resources such as laboratory equipment, materials, space, etc. However, in practice, resources seem to be less available for women (Gupta and Sharma 2003). For instance, women faculty at times face problems in obtaining research scholars. Further, while student choice of an advisor is considered for allotting advisors in most elite institutes, the weightage given to their choice varies from institute to institute, and even department to department. Both student prejudices and department policies could marginalize women faculty. Women are usually junior and are seen by the students as having less time and low contacts (Gupta and Sharma 2003). In a study of doctoral students at IITs (Gupta 2007), about one-third of the male doctoral students believed that men are better teachers and researchers, while only about one-tenth of the women thought so. A student's choice of advisor is considered in the CSIR labs also. A woman, Scientist C in a CSIR lab, explained that the requirement for students in their institute was more than the intake, and students are given a choice to join any faculty. So, some faculty might have none. 'There is domination of seniors in research; if you are in a committee, you favor yourself; there is no transparency; they have better access to instruments; they take students; students go to the faculty that already has students' (Gupta 2013: 100). In some institutes/universities where departments play a major role in the decision on who takes which scholar, such decisions are usually influenced by the seniors, and politics and favoritism play a role (Gupta and Sharma 2003).

Further, the names of departmental committee members are usually decided by the seniors who are usually men (Gupta 2016). Men usually appoint men as committee heads (Gupta 2013), suggesting 'homosocial' reproduction (Connell 2000), which is seen as central to gender inequality in academia (Cohen et al. 2023). There are various other practices suggesting how men tend to promote men. This happens, for instance, through application of a gendered decorum of behavior that marginalizes women. Thus, it is supposed to be womanly to be submissive (Mukhopadhyay and Seymour 1994). As a woman scientist stated:

> Some men colleagues initially wanted to force their views on me; they thought that being a woman, I should simply agree to their viewpoint.
>
> *(Gupta and Sharma 2002: 908).*

The problem is greater in the traditional engineering departments, or in those departments that have relatively older faculty members, or in those with very few women (Gupta and Sharma 2002).

The practice of science requires dealing with the administrative staff. An inefficient system in many institutes, ridden with red-tape, and a hierarchical mindset combine with gender biases to adversely impact professional careers of women scientists. For instance, in universities, appointments as external examiners are decided at the lower administrative level, and these appointments are a source of greater visibility to the faculty in their discipline. Subrahmanyan (2009) found that women faculty were not usually appointed as external examiners because women do not fraternize with the staff (mostly men), while male faculty would go to them and ask their name to be put down for it. A recurring theme in a study of a CSIR labs was that the CSIR system is slow moving and that contacts and a high position in the hierarchy ease the pressures of bureaucratic procedures. As Manish (male, Scientist G, CSIR2) commented:

> … lethargy of the system is there at all levels [in] procurement, recruitment, getting funds, paperwork … [a] person who is well connected, things move for him.
>
> *(Gupta 2016: 448).*

The term 'well-connected' here implies contacts with those in higher positions. Interpretation of rules might vary, and the administrative staff, which usually implements the rules, tends to favor those with a higher status in the institute's hierarchy. For instance, according to Jagdish (male, Scientist C, CSIR1):

> The administrative staff is not very helpful. Their help depends on what kind of relation you have with them or how senior you are; … Show me the face I will show you the rules.
>
> *(Gupta 2016: 449)*

As stated earlier, there are few women in higher positions. Also, routine procedural matters of purchase and funding involve non-scientific staff. Women scientists of junior rank face problems that combine rank with gender. For instance, according to Rukmini (female, Scientist C, CSIR1):

> Sometimes it is very difficult to speak to a male staff member, they feel offended ... lower staff is very adamant, they will not listen to you because of your [younger] age and gender.
>
> *(Gupta 2016: 449).*

Although research in leading institutes is well-funded, obtaining funds for seminars and conferences might also involve bureaucratic hurdles, and some women scientists in academia (Gupta and Sharma 2003) might not apply for funds at all because the application requires a great deal of 'running around' and decisions suffer from favoritism. A respondent in this study (Gupta and Sharma 2003) overheard a clerical staff member (male) saying 'lagaane do isko chakkar' ('let her keep running around'). Sabharwal et al. (2020) show the discrepancy between the proportion of women faculty at an institute and the women attending conferences; that men academics attend proportionally more conferences than women academics. Such conference participation inequalities directly impact the indices related to professional development.

The scientific workplace as essentially 'male' is exemplified in the concept of an ideal scientist/worker.

Gender, 'scientist', and 'ideal worker'

There is a pervasive stereotype of a 'scientist' as a man and a norm of 'male model' in science. Women must conform to this norm, which requires, for instance, devoting long hours to lab or research (Etzkowitz et al. 2000). This overlaps with the notion of a universal/ideal worker in organizations, which is a male unencumbered by familial responsibilities (Acker 1990; Brumley 2014; Collinson and Hearn 1996). It is premised on the ideology of separate spheres with the public sphere as the man's domain.

The masculine norm of long working hours is prevalent in the Indian research institutes reflecting the notion of a 'male model' of science. This is evident in the following statement of a male scientist (Scientist F) at a CSIR lab (Gupta 2016: 453):

> A capable scientist is one who has more recognition. Actually, the *time period* that [a] man can devote is more because a woman cannot. This affects her work and it takes a long time to complete the same work.

This statement highlights the relationship between the notion of time required for doing 'good science' and gaining success. Further, since women are perceived as devoting less time, their research is assumed to be below par as reflected in the following statement of a senior scientist (Scientist G) at another CSIR lab (Gupta 2016: 453):

> Research of women is *mediocre*; you need to be immersed in research. Women are *family oriented* because in India it's given that women sacrifice more.

That women's 'research is mediocre' was probably a generalization based on a limited sample of two women scientists in his department who happened to be average researchers. Thus, women's failures are magnified due to a 'heightened visibility'. Also, there is a lack of recognition for women's achievements. The belief that the scientific research of women is not at par with that of men is negated by the evidence (Gupta et al. 1999; Hasan, Sharma, Khilnani, et al. 2012).

Labeling entire women's research as 'mediocre' (as in the above statement) is a glaring example of an underestimation of women's performances, also referred to as 'professional diminution' or questioning a woman faculty member's professional authority, which is one of the most common micro-inequities in the academic profession (Benokraitis 1998: 8). In a study of academic science institutes (Gupta and Sharma 2002), only about 34% of women academics could say that their male colleagues do not underestimate women's work. This diminution is further brought out in the statements of women scientists that 'men colleagues take a long time to see that a woman is equally competent'; 'men are unable to accept that a woman of the same designation can do better than them' (Gupta and Sharma 2003: 607).

The underestimation of women and the resulting condescension are reflected in men faculty members being sometimes too appreciative of the small achievements of women faculty (Gupta and Sharma 2003: 608): 'Even if I do something of little merit, men faculty show a lot of appreciation' (Dr Savita Jain, Assistant Professor, physics), or 'The men colleagues are extra polite towards us which is quite irritating' (Dr Sheela Roy, Assistant Professor, electrical engineering). A woman faculty member in a Subrahmanyan (2009: 188) study recounted how her HOD, when convening a committee for administrative purposes that had no man in it, would ask her, 'Can you manage?'; or in another instance, 'You are all ladies. Shall I put a man along with you?' This reflects diminution and a sense of male entitlement at the scientific workplace that overlaps with gender illiteracy. As remarked by a woman professor of physical sciences at Jawaharlal Nehru University:

> More often than not, men don't even realize that something they have said or done was offensive or threatening to women colleagues or students.
> *(quoted in Masoodi 2016).*

That women scientists devote less time to research compared to men might also be a myth. A survey conducted in 2010 on 312 women and 161 men in research by the Indian Academy of Science, along with the National Institute of Advanced Studies, showed that women often work harder than men, with 62.8% of women spending 40–60 hours a week at work as compared to 46% of men. A higher proportion of men (25.5%) work less than 40 hours compared to women (16.5%); it is in the category of working extremely long hours of 60 hours or more, that the proportion of men exceeded (at 28.6%) compared to 20.5% women (Kurup et al. 2010: Table 41).

Because of the dual burden assumptions, 'men feel that women are less committed ... [so that] they [women] have to be two and a half times better than men to achieve similar recognition' (Gupta and Sharma 2003: 607). Professional diminution is greater at the initial stage of a career, so that the time required to prove that they are equally capable may be quite long.

> The colleagues initially questioned my professional competence and tried to put me down in meetings and general talks. I had to be on my guard. [A] lot of professional energy was dissipated in that, otherwise I could have done much better (Dr Rita Gupta, Associate Professor).
>
> *(Gupta and Sharma 2003: 607).*

The notion of a scientist as a male encompasses not only a 'male model' but also a perception of science/engineering as being more suitable for men than for women. For instance, studies on elite Indian academic institutes have found stereotypes such as, women would find it difficult to handle field work, women cannot teach boys effectively (Gupta and Sharma 2002), women are incapable of handling a certain engineering section, as seen in this statement of a woman faculty (Gupta and Sharma 2003: 613):

> A problem occurred when I was given the charge of the PQR engineering section in our department four years ago. Some seniors thought that a woman could not run the section and raised the issue in the departmental meeting. Though the section has continued to be under my charge, the rules related to its running have been flouted. Till date [sic], I have not been given peons and technical assistants.

The concept of an 'ideal worker' (such as the scientist as an employee in an organization) as always available and visible for the organization is a male (Lewis and Humbert 2010). This is manifested in meetings scheduled beyond the office hours, which affects women's participation due to work-life balance issues (Gupta 2016; 2017). However, the notion of a man as an 'ideal worker' remains unimpaired when men alter practices to create work-life balance. Men wanting to pick up kids from daycare might lead to rescheduling of the meeting time.[1] The primacy of men in the public sphere is further reflected in the contradictory assumptions about similar actions of men and women. For instance, 'if I am late in coming to the department, it is attributed to domestic duties at home; while if a man is late, he is considered to be busy in some legitimate work' (A woman scientist in Gupta and Sharma 2003: 613). Women scientists are seen as negating the ideal/universal scientist concept of devoting long hours to the lab and being available to the organization. The concept 'excludes and marginalizes women who cannot ... achieve the qualities of a real worker because to do so is to become like a man' (Acker 1990: 150).

The contradiction in assumptions becomes even more glaring when women scientists are regarded as 'privileged'.

Women as 'privileged'

As mentioned in chapter four, Gupta (2007; 2010) found that male doctoral students believed that female doctoral students were privileged in terms of obtaining greater cooperation, being graded softer by faculty members, or being given undue favors. They were overwhelmingly unaware of women's problems relating to interaction, prejudices, and lack of academic help.

The notion of women's 'privilege' carries over to the faculty level. Government schemes convey the impression that women get 'things' easily, for instance, in a study (Gupta 2020: 699), Prashant, (male, Scientist F, CSIR1) said, 'Actually, women get more opportunities and get things more easily than men ... because government is encouraging them ... if there is some award, women being few at high positions, face less competition and get it easily'.

Women scientists are perceived as having a less rigorous professional life and, therefore, advantaged. For instance, Manish (male, CSIR2, Scientist G) in the Gupta (2020: 698) study, said, 'There is no bias here. On the contrary, women are *privileged*. They come late [to the lab], go back early'. Some men scientists perceived a bias *against men* since men were given more responsibilities, as, for example, when women were 'let off' certain tasks (Gupta 2020: 699). However, more responsibilities for men also imply denial of opportunities for women, for instance, in being 'let off' the charge/membership of a departmental committee. In perceiving women as privileged, 'othering' of women scientists occurs that marginalizes women scientists.

Masculinity and caste

Dalit/Adivasi women form a minority within the minority. The exclusion, discrimination, and denial of opportunities are greater for them as they are at the bottom of the social hierarchy. Overall, in 2020, less than 1% of professors (male and female) came from the Dalit/Adivasi communities at the top-ranked institutes among the 23 Indian Institutes of Technology (IITs), according to data provided to the journal *Nature* under right-to-information requests (Paliwal 2023). Rao (2013) found that Dalit teachers are discriminated in an allotment of courses and students, which affect career advancement. Access to, and participation in, conferences is proportionally lower for Dalit academics than for men and upper-caste academics (Sabharwal et al. 2020). In 2021, an Assistant Professor at IIT Madras resigned, stating caste-based discrimination at the hands of his colleagues as the reason behind his resignation.[2]

Women scientists of lower castes face issues that magnify the consequences of masculinity when it intersects with upper caste institutions and marginalizes them even further in the scientific community. In a study on scientists at CSIR labs (Gupta 2013), Sanjana, a woman scientist belonging to a Scheduled Caste, Scientist C at CSIR2, noted (p. 100–101):

> Even though I had credentials and had more papers than the other person who was taken under open category, I was not taken [recruited] under open category but under reserved category and given the position of Scientist B
>
> She feels quite bitter about this initial appointment saying, 'Scientist B was not where I should have joined; we are forced to write "caste" on application forms and then denied what should be given to us as an equal' (p. 101).

This affirms Rao (2013) that various policies and practices that 'label' the communities in institutes are stigma-based categorizations and are detrimental to the success of Dalits/Adivasis.

After Sanjana's promotion to Scientist C, when she became independent, she started writing projects for funding. 'I applied many times to agencies outside this lab, but failed to get any project. I have been called for presentation but was not given any reason for their refusal. However, I have been working on a couple of joint projects and have one from this institute' (Gupta 2013: 101). She has also failed to get PhD students until now, without which research is difficult. She has project trainees but they are supposed to be employed only for six months. She has not been part of any committees. No one has approached her for this. She does not see it as a caste issue, but feels that 'men tend to involve men more in committees and meetings; women don't say that they want to be a part of it' (p. 101). Thus, the problems faced by women scientists of lower castes reveal an overlap of caste, gender, and systemic inefficiencies.

Masculine culture at private sector research labs: a case study

The private research labs provide an example of how specific exigencies of a profession, along with the generally prevalent socio-cultural norms (such as hierarchy), shape masculine culture in organizations and impact the status of women professionals. In the two pharmaceutical labs studied (Gupta 2017),[3] the scientists were mostly PhDs or doing a PhD. The research problems of the scientists are decided by the pharma companies, which also maintain control over data, materials, tools, and results. They prefer to patent new findings rather than disseminate them in a journal. Research in these labs is governed by commercial interests. The scientists work on the target-oriented modules set by the management. Delivery of timelines as per these targets is of prime importance.

The project manager, along with the company bosses, decides the timelines for a research project and the team has to work accordingly. The actual working is left to the scientists in the team. The team is composed of scientists in a hierarchical arrangement from junior to senior scientists. In both types of research labs, there was not only a preponderance of men in numbers, but men occupied most of the higher positions reflecting greater formal power.

In these private research labs, an 'ideal worker' was one who was visible and available, as elsewhere (Lewis and Humbert 2010). An informal practice of judging employees' contributions in terms of hours spent at work irrespective of the amount or quality of work accomplished was widely prevalent, which affected women adversely (Gupta 2017). As Jhanvi (female, junior scientist in Private1) remarked:

> When a man and a woman are at the same level but only one has to be marked for promotion, the superiors see who has worked hard and this is judged by the number of hours given to work ... I may finish my day's task and still go back on time. But this is not recognized by the superiors.
> *(Gupta 2017: 263)*

The culture of working long-hours is aggravated in the Indian hierarchical cultural context. If the boss stayed late, the juniors were unable to leave early, reflecting a high power-distance. A hierarchical culture in India heightens the need for visibility for the benefit of bosses. Visibility is required not only in terms of long hours but also in the form of interpersonal interaction and rapport with the seniors. Interaction norms that inhibit networking with men, and meetings kept beyond office hours (sometimes without much agenda and, therefore, wasteful), impacted women in these labs also. Informal contacts and networks for women scientists in the private labs were further restricted by the inflexibility of the work schedule that required presence at the workplace in the office hours and by gendered practices. An example of the latter includes the requirement for plant visits in pharma labs. These visits provided networking opportunities, but they were structurally unfriendly and unsuitable for women (Gupta 2017).

In the private research labs, companies have certain preferences in recruitment as well as during promotions that act as an advantage for men. For instance, in this study (Gupta 2017), more value/preference was attached to the worker who had gained experience in similar companies. Constrained by family, women most often did not change companies as it usually required relocation to a different city. The preference for experience in other companies was not a part of the formally written rules but practiced informally. It led to a lower position for women in the company than a man who had the same experience as her, but might have worked at various other companies. Such informal practices disadvantaging women exemplify 'gliding segregation' (Holt and Lewis 2009), that is, the process of women and men starting at the same place with the same level of education but ending up with different opportunities for development due to gender. Plant visits, as mentioned above, were important for promotions also, but extremely difficult for women due to lack of facilities. This exemplifies how organizational 'gender neutrality' translates to facilities favoring only men.

Further, the women in the most male-dominated professions struggle with similar socio-cultural norms and practices that devalue their status at the workplace. These are discussed with reference to women in the academia and public/private research labs below.

II. Construction of female disadvantage

The idea of women's primary domestic role creates assumptions, beliefs, and practices in the work environment that tend to disadvantage women and devalue their status at the workplace Such beliefs and practices include:

a. The dual burden assumptions of male colleagues.
b. Biased treatment in pregnancy.
c. Inadequacies of work-life balance policies and structures.
d. Gendered assumptions about leadership roles.

Assumptions of men about dual burden

An ideological belief that women are the center of the domestic sphere is coupled in the recent decades with the growth of Hindu nationalism, a consequent reassertion of the notion of family, and family values as the fulcrum of Hindu identity among the urban middle classes (Radhakrishnan 2011). Studies elsewhere show that while being 'flexible', or flexible to the needs of the workplace, is valued highly, women are not looked upon as 'flexible' workers (Holt and Lewis 2009). This section analyzes how assumptions of women's family roles result in loss of opportunities and compromise women's positions in the organizational hierarchy. Although many examples resonate with studies in other parts of the world, they also go beyond those studies to indicate a more entrenched belief in the separation of spheres.

The belief in separation of spheres is primarily manifested through the assumptions of men that women are constrained at work due to their devotion to their family. For instance, women faculties at national institutes have reported facing interview committees for appointment that ask gender-based questions (Gupta and Sharma 2002), such as, 'why do you want a job?' or 'how will you take care of your family?' The interview committees (for the appointment of faculty) reflect various social prejudices, such as the fear of an unmarried woman candidate getting married and losing commitment or of a married woman leaving the job to join her husband working elsewhere. In a study on the CSIR labs (Gupta 2016: 453), a junior woman scientist (E1, CSIR2) said, 'Awareness of a dual burden on women affects hiring. The woman has to prove that she is better than a man for any job'. A senior male scientist, also an HOD at a CSIR lab, on the one hand, remarked, 'whosoever is qualified should be appointed', but on being asked what if both male and female candidates are equal, he replied, 'the male is preferred because he can work for longer hours as compared to a female' (p. 453). The scientists featured on 'The Life of Science' reported being asked personal questions during job interviews like 'you have applied for a job in this city but your husband is in another city or country, so how will you manage?' (Thite 2017). There are stereotypes that a woman, by being appointed as a faculty member, is taking away a man's seat (Gupta and Sharma 2003). As a result, a woman's bio-data has to be 'really good' and 'women have to work harder to prove themselves' (Gupta and Sharma 2003). This is further exemplified in the UNESCO study (2022), which found that a higher proportion of women faculty had received merit-based scholarships during their education than men.

The assumption of women's domestic roles as primary is often used as an instrument to promote a male candidate, which isolates women scientists in organizations from important tasks/committees (even though a woman might be willing or capable of handling). For instance, a senior woman scientist at a CSIR lab remarked (Gupta 2016: 452):

> Whenever something important is going on in the division and someone has to take charge of it, they automatically think of men. For a long time, I was the only woman in this division. It was assumed that I have a lot of work at home. In the guise of helping, they sideline.

This exclusion often takes the form of 'letting off', that is, allowing a woman colleague to opt out of important tasks, usually administrative and non-academic, requiring long hours. A woman is not always consulted for those tasks as apparent from the above statement. At times women themselves request being let off. However, such requests might imply structural constraints and lack of support (Gupta 2020). This 'letting off' assumes that family role of women is part of the 'Indian culture'. There is also a lack of confidence in the abilities of a woman. For example, in the same study, a senior male scientist said:

> For a particular task, if [we are] selecting people, you are avoiding women, maybe you have more confidence in men, or [a] woman may have to work at odd hours or have small kids, so it's a human thing that, o.k., go early, it's our Indian culture ...
>
> *(Gupta 2020: 696).*

This identification of women's family roles with 'Indian culture' masks the glass ceiling effects. The use of dual burden assumptions by men colleagues for excluding women is reported by various women scientists, for instance: 'men want to make men as committee heads because men say that a woman would spend more time on the domestic front'; 'sometimes opportunities suffer because they [men colleagues] calculate that women can't devote time, only when there is no option, opportunities are given to women' (Gupta 2020: 697). These opportunities are participation in seminars, consultancy, in any difficult work, in going abroad, or taking part in decisions. Hence, on the one hand, there is an assumption that a dual burden on women renders them unavailable for significant tasks; and on the other hand, there is a sense among women scientists of being left out. While there are women scientists who themselves do not wish for an administrative position or committee headships, but as a woman scientist at a CSIR lab commented (Gupta 2016: 453): '70% of women will not go after leadership position[s], but if it comes to them, they will not say "no" ... Men lobby for it ...'

Thus, the repertoire of 'Indian culture' in terms of assumptions of women's devotion to family enacts a gendered system in organizations through 'automatic sex-categorization' (Ridgeway 2009). This assumption in India is so strong that a woman committing extra hours to work might be perceived by the colleagues as compromising her primary domestic role. For instance, a woman scientist overheard a male colleague saying 'she always comes at nine in the morning, she does not seem to have any work at home', and she felt that it was expected of her to go home at five in the evening (Gupta and Sharma 2003: 613).

In the study of private labs (Gupta 2017), dual burden perceptions of seniors affected promotions of women scientists. The latter were often given less important tasks. They were denied promotions because of the perception that as mothers they might not be able to handle the enhanced duties or if a promotion required mobility or transfer, her non-availability was assumed *without asking* her.

Biased treatment in pregnancy

The masculine image of a scientist is incompatible with a woman scientist, even more so when she is pregnant. Martin (1990) suggests that in the male-dominated organizations, a pregnant employee is an alien. This is reflected in the situations faced by the Indian women scientists, as for instance, Poonam, Scientist F in CSIR2 said:

> [During my] first time [pregnancy], my seniors troubled me by giving difficult job; second time, they ignored me, they thought that I should go home, why am I coming here, my husband is earning, that two years she will not do any work. They used to tell me that I should leave the job and go home.
>
> *(Gupta 2013: 59)*

The feeling of hostility for a pregnant woman scientist occurs because pregnancy seems to be an intrusion of the 'private' (or domestic) sphere in the 'public' (or economic) domain. Socialized in the Indian patriarchal system, many women believe that family/childcare issues are women's problems. For instance, justifying the negative feelings of the male colleagues, CSIR1, Richa, scientist G, said:

> In pregnancy, *thoda sa to dekhiye hota hai, ab chhuti pe chhalee jayenge* ('see, colleagues do have some negative feelings, they feel you would go on a holiday'). But one has to understand their state of mind. I will feel the same way. I will also tell my student-*jaldi kar lo nahin to phir chhe maheenein nahin kar paogi* ('do it fast, otherwise you will not be able to do it for the next six months').
>
> *(Gupta 2013: 60)*

However, there is a subtle difference in male colleagues reflecting on pregnancy as a holiday and a woman supervisor persuading her student to try to finish some work. The first one is a belief that pregnancy is a systemic loss, while in the latter, the student herself would be a loser.

In the private labs, the bosses might be considerate during the pregnancy of a woman scientist, but this has its own price (Gupta 2017). If a woman is pregnant, bosses usually might not allow her to do heavy work and would give her a reduced workload. However, the work diminishes her career, and even after her return from the pregnancy leave, she might not be given important work. The male colleagues in the team might be hostile. The women scientists reported that 'the attitude of the colleagues in pregnancy is not very much positive. Some kind of resentment is there' (junior scientist in Gupta 2017: 269). Pregnancy leave is seen as 'time off', as, for instance, in the statement, 'women can take off for more than four months' (male scientist in Gupta 2017: 269). The statement reflects a lack of respect for women and indicates how women's biology is seen to negate the notion of 'ideal worker' or a full-time, available, and visible worker, which devalues them in the public sphere.

Inadequacies of work-life balance policies and structures

Work-life balance structures, while actually aiding work-life balance, can play a major part in diluting the male perception that women find their work role difficult to manage along with their domestic role. However, typically, organizations in India lack childcare and daycare infrastructure compared to the West. It is becoming more common mostly in response to globalization and the need to attract and retain the best talent to the organization but not because of an intrinsic belief in them (Gupta 2017; Rajadhyaksha 2012). Maternity leave and other leave benefits are provided more in the government labs than in the private labs, as, for example, a childcare leave for women up to two years (recently extended to single men with children), the Leave Travel Concession (LTC) and sabbatical leave of one year for advanced study or research are provided in the government institutes/labs but usually not in the private sector. Some organizations (public and private sector) have better work-life balance policies and infrastructure than others.

In practice, the work-life balance policies seem to contribute to hegemonic masculinity at the workplace. As mentioned earlier, these policies are treated as 'privileges' for women by the male employees and the pregnancy period is often seen as 'time-off'. Men rarely avail themselves of flexible timings offered in the private sector. According to Rajan (male), principal scientist at a private lab, 'organization has flexi-time policy in which one can come late [at 11 a.m.] and go back early. I have availed it only once or twice [due to family requirement]. It is basically for women. Women avail of it more' (Gupta 2017: 267). This study found that those who use such flexi-timings are viewed unfavorably; by implication, men are viewed as more regular and, therefore, more committed to work. Although work-life balance policies have their strengths, researchers have shown that they fail to tackle the male model of working (Lewis et al. 2007), and the structures of patriarchy that necessitate such interventions (Rajadhyaksha 2012). The perception that such policies are for women reflects both the masculine notion of worker as a full-time committed worker and women's centrality in the domestic sphere, strengthening the idea of separation of spheres.

Pregnancy period results in career loss to a woman. In academia/research, the period lost to a woman in pregnancy lowers her research publications and reduces chances of promotions along with her batch of recruits (Ananth 2014). In private organizations, at the time of promotions, a woman who took pregnancy leave is compared to a man who has worked for the entire 12 months. The pregnancy leave period deprives women of career growth in an important phase of their life when they are still at the junior level (Gupta 2017).

Company/organization considers the pregnancy period as a loss to itself. This was recently witnessed in the reaction to an amendment in 2017 to the Maternity Benefit Act of 1961 that enhanced and brought maternity leave of the public and private sector at par (to 26 weeks). However, many companies, both in the government and private sector, were reluctant to hire pregnant women due to this extension of maternity leave, and some even began firing women employees. Hence, the government had to intervene and announced that it will give employers the salaries for seven of the 26 weeks of maternity leave for women earning more than Rs 15,000 a month.[4] As per an amendment to the maternity leave rules in 2017, employers with 50 employees have to provide crèche facilities.[5] While such facilities have yet to be incorporated at all such places, their quality remains unregulated; the daycare centers for grown-up children are rare and elderly care facilities remain poor.

Can women manage the dual burden and do well professionally? Xie and Shauman (1998) found both men and women married scientists to have significantly higher (by 7–11%) rates of productivity than unmarried scientists. Research from various parts of the world indicates contradictory evidences on the relationship between motherhood and scientific productivity. Some research shows that motherhood could have a positive effect on productivity (Fox and Faver 1985). Still, others have found that married women with children have lower research productivity (Hargens et al. 1978). According to Stack (2004), only women academics with young children publish less than other women with children; academia's flexi-time facilitates balancing childcare and academic work.

In India, research regarding the impact of marriage and motherhood on women's research productivity is negligible. In most of the elite academic institutes and the government research labs, there is considerable freedom to the scientists to decide their work schedule. Hence, the issue is not of flexi-hours but of time management for women in such organizations. For those who have grown-up children, there are a different set of problems compared to those with younger children. As a woman scientist in a CSIR lab remarked, 'Flexi-hours is there. But it is horrible in the first two to three years when kids are small. Even now if an elderly at home is not well, you have to take care' (Gupta 2013: 57). Women in institutes lacking in campus residence for scientists have greater time management issues due to commuting. Also, in such cases, the presence of a crèche alone is insufficient as a daycare is required for school-going children.

Responsibilities of childcare and care for elders are often the primary reason for the breaks in career (Kurup et al. 2010; Agashe at al. 2022). Such breaks tend to lower women's position in science as they get left behind in research and in organizational hierarchy. Family roles have an impact on opportunities for women in terms of fellowships abroad, positions as visiting scientists, and traveling for conferences and workshops (Gupta and Sharma 2002).

In the absence of infrastructural facilities, women receive assistance from non-institutional sources (spouse, in-laws, parents, paid help) for childcare (Kurup et al. 2010). Such sources of assistance, such as the parents-in-laws, have their own complications arising from gendered expectations (Gupta and Sharma 2002). The support of a spouse in professional matters and his help in domestic roles have been critical for a majority of women scientists in this study (Gupta and Sharma 2002), which illustrates how individual families might be changing.

However, the working status of a woman fails to create a major impact on gender-role ideology in the families. Research on work-family conflict (Rajadhyaksha and Velgach 2015) of working women in India found that although the working status tends to create more egalitarian gender-role ideology (gender-role ideology or GRI, refers to an individual's attitudes and beliefs about the proper roles of men and women), *working women seem to lean towards a more egalitarian GRI relative to working men.* Higher family interference in work roles and higher mean levels of childcare hours, elder care hours, and housework hours for women suggested that men may be taking on family roles at a slower rate than women may be embracing a work identity. Thus, as authors argue, this creates the rationale for *greater institutional rather than family-based support* for family demands, such as childcare centers, eldercare centers, assisted living, and hospice care.

Careers in scientific research seem unfriendly to women through another statistic. There is a higher proportion of 'single' women than men in science. In a survey among the men in research, only 2.5% were unmarried; among women in research, those who never married constituted about 14% (Kurup et al. 2010), that is, about 5.5 times the men. Single women scientists (including divorced/widowed/never married) do not necessarily demonstrate higher research activity than married women in terms of research publications and participation in conferences. For a single woman, social isolation and higher difficulties in collaboration and interaction with men colleagues possibly lead to being a less prolific researcher than a married woman (Gupta and Sharma 2002). This is also found in research in other countries (Aiston and Jung 2015), and the authors suggest that family-related variables are overrated in their relationship to productivity.

A gendered work environment and multiple role management, absence of work-life balance structures, efforts at time management, and self-discipline extract their cost leading to exhaustion for women scientists, at least physically at the end of the day (Gupta and Sharma 2002). Women scientists have to work twice as hard as men due to workplace and institutional barriers (Gupta and Sharma 2002; 2003). In attempting to balance multiple roles, they compromise on opportunities and are unable to use their potential fully, which forces women to scale down their expectations from their career.

Despite a lack of facilities for work-life balance, it is essential to note here that the publication productivity of women does not vary substantially from men in most fields (as shown in the earlier chapter). Research activity of married women is higher than that of single women (Gupta and Sharma 2002). Even if research productivity of women is less than men, as in the US, for instance, it is mainly due to systemic factors. The focus on familial burdens of women takes away the focus from the systemic and organizational biases that impact women's positions in science. Absence of work-life balance structures and policies is both the cause and consequence of the gendered expectations. A dual burden on women scientists becomes significant mainly due to the gendered expectations and systemic lacunae that discourage work-life balance. As Padavic et al. (2019) remark that the work-family explanation has become a 'hegemonic narrative' or a pervasive, status-quo preserving story that prevails despite countervailing evidence.

The 'rationality' and gender-neutrality of institutions mask masculinity of structures. This is evident not only in the lack of work-life balance structures, but also in the lack of basic provisions for women. A study at the beginning of the millennium (Gupta and Sharma 2002) noted an insufficient number of separate washrooms and their poor maintenance and an absence of on-campus childcare facilities in the elite institutes. Even in recent times, lack of ladies' washrooms and their shabby conditions in most scientific institutes have been noted by Dogra and Jayaraj (2023) in their journey across various labs. Feeding stations continue to be mostly absent. The lack of gender-sensitive facilities has a negative impact on women (GATI report 2022). Unsupportive infrastructure constrains careers and implies women's secondary status in science.

Leadership and gender

While there are various challenges to women in junior positions, as discussed previously in the chapter, those women who do manage to counter barriers and move ahead in the organizational hierarchy encounter further obstacles in leadership positions.

Leadership is associated in South Asian countries, such as India, with particular types of masculinity, competitiveness, ruthlessness, and politically networked (Morley and Crossouard 2015). Women do not fit into the gendered leadership style expectation and face hostility if they are agentic, which is seen as a masculine style (GATI Report 2022). There are stereotypes 'because of which they [men colleagues] would like to keep women away from administrative responsibilities. They would say, "that poor thing; she would have no idea how crooked people could be" and start patronizing' (male scientist in Gupta 2020: 700). Socialized to believe that women are obedient and docile, not only men, but some women scientists also believe that women are less capable of managing or are softer in making decisions compared to men (Gupta 2017).

Leadership is not only masculine but is also visualized as a 'male' domain. There are constraints in giving higher positions to women, 'it is in the minds of people that if she is a mother, she can't handle' (female scientist in Gupta 2017: 266). There is a lack of respect for the leadership abilities of women, and male colleagues find it difficult to work under a female boss (Gupta 2016). Old boys' clubs often isolate women from leadership chances (GATI Report 2022). Senior women academics who are eligible and willing to accept the leadership position might be denied (Gupta and Sharma 2003).

The reasons given by the women scientists for a lack of women deans in the Gupta and Sharma (2003) study were: there being very few women faculty at the senior level; gendered notions in the institute such as, a respondent was told that women could not work at night; or that 'men do not like to be bossed around by a woman'; or that the 'institute will not prefer women deans'. Further, such positions require politicking and *chamchagiri* (ingratiation), which women academics tend to avoid. The other reasons stated by women academics included a lack of ambition on the part of women, family constraints, and lack of time—reasons which are also a reflection of successful internalization of gendered norms by women (Gupta and Sharma 2003). Most women scientists will not lobby for leadership positions (Gupta 2016). An ideal scientist is seen to work for science and not for recognition or positions; hence, a woman's aggressive demand for a leadership position would render her non-scientific and non-feminine, unfit for the role (Arora 2022).

Women bosses have to grapple with the stereotyped assumptions of those junior to them. In a study of private labs, a woman scientist at the second-most senior level in her R&D division (Private2) recounted how a new male entrant, a junior scientist, had apprehensions that having a woman for a boss, she would devote fewer office hours and so would give away all the work to others so that they would be overburdened (Gupta 2017). She remarked that the earlier group she headed consisted mainly of women and she never came across such apprehensions.

Gendered norms of mixed-sex interaction might become a tool for envious colleagues to downgrade a woman in a leadership position. For instance, a woman academic who was a Dean at the Banaras Hindu University, found the initial Vice-Chancellor supportive and 'work was going really fast for me (sic). If I needed to see him, I would call his office and if he was busy, he got back to me later' (quoted online in Jayaraj 2017). But when the Vice-Chancellor changed, it emerged that some people had been very envious of her. 'I got some letters that claimed I had an affair with the previous Vice-Chancellor' (Jayaraj 2017). Thus, norms against mixed interaction could be easily misused to damage a woman's reputation.

Some women scientists that move up in leadership positions might adopt 'masculine' behavior, such as being aggressive. Female leaders often face the paradox of whether to emulate a masculine leadership style or a stereotypically warm and nurturing feminine style. In the first case, their male subordinates will dislike them; and in the second, they will be liked, but not respected (Kawakami et al. 2000). A male scientist in a CSIR lab remarked that women in higher administrative positions are harsh to junior women scientists (Gupta 2013). A woman scientist in a senior position in a private lab had noticed that when women come to power, some of them adopt very aggressive behaviors, 'probably because they had struggled so much in the process of rising, which is tough in corporate, [that] they have built up their defences; becoming offensive is a part of that defence' (Gupta 2017: 269). Through aggressive behaviors, women fulfill the masculine gender role script in leadership work, and end up 'doing gender' (Kelan 2010).

Denial of gender inequality as systemic

While not denying the existence of gender inequality in numbers and positions, the scientists (male and female) often refuse to blame their lab/organization for inequalities by delinking gender issues from the organizational domain, as found in a study (Gupta 2020) on Indian research labs. This delinking occurs by separating individual incidents of discrimination from the 'system', and rationalizing gender inequality as 'social'.

Interpretation of overt discrimination as 'individual'/unrelated to system

Women scientists at times encounter not only covert but also overt discrimination in day-to-day interaction, which might affect their careers. However, such incidents of discrimination are seen as isolated incidents and blamed on individuals (Gupta 2020; 2016). They are not seen as organizational discrimination and add to the masculine nature of the work environment. For instance, a woman scientist (E2) at CSIR1 (Manya), who was the only woman in her department for 15 years, was denied promotion due to biased reports of her HOD.

She said (Gupta 2020: 697):

> My institute, my director, and most of my colleagues are lovely. Gender bias occurs usually because of one person. It occurred because of the bad mentality of my HOD. Then there are two other colleagues. HOD doesn't publish. He is an HOD by age/seniority. Scientifically he is very weak. So, there is professional jealousy ...

She believed that gender equality 'is respect, which I am not getting from my male colleagues' and that there was gender equality in her organization. '... my gender discrimination is personal, it is nothing professional'. Another woman scientist, Maansi, in the same study (Gupta 2020), who joined as a Scientist B (CSIR2) in 1987, was not independent and had to work under the supervision of the HOD. She encountered serious difficulties from her HOD until he retired in 2005. However, when asked if she thought there was gender equality, she said 'yes' because 'the system provides equal opportunities to both men and women'. Further, she remarked, 'I have patience and mental strength; hence, I was able to fight him' (p. 697).

Although noted as discrimination by individuals, the discrimination faced by Manya and Maansi was organizational. The perpetrators belonged to the same organization as the victims and held senior positions in the system. In Manya's case, a few men colleagues were colluding with the HOD while some others were mute spectators. Maansi hinted at systemic fault for not removing the erring individuals when she remarked that 'If middle manager is notorious, he should be removed' (Gupta 2020: 697). Thus, gender biases are widespread and Manya had to fight for even the basic facilities, such as lab space. Further, the discrimination occurred over a long period of time and could have recurred if the discriminating HODs were still around. Yet, Manya and Maansi perceived no discrimination in the organizations since they did not recognize any problem with the 'system', which as Maansi said, provides 'equal opportunities'. While discriminatory individuals or men as a group are held responsible, the system, by having no provision to take corrective action against such individuals, ignores individual discrimination.

The above incidents indicate that the system is equated with formal rules, and discrimination is seen as individual incidents. In perceiving the individual incidents of discrimination independent of the system, the system is perceived as rational and gender neutral. Further, the system is equated with formal rules but not with the organizational practices. As a result, biases in meetings, selections, and day-to-day interaction are seen as processes unrelated to the formal system. Finally, a lack of critical mass creates a climate of acceptance of male practices and leaves little scope for dissent.

In constructing discriminatory incidents as individual, women exhibit 'individualization' in the Western sense that individuals assume responsibility for their life and for all their choices (Beck 2000). However, this process of 'individualization', a part of the global process of 'reflexive modernization', overlaps with traditional social forms (such as family) in post-colonial societies (Belliappa 2013). This is evident in the dual burden assumptions that construct gender inequality as 'social'.

Gender inequality as 'social': self-blame and humiliation

Often women as a group are believed not to be doing enough professionally due to their domestic role, since the latter is mandated by the social structure (Gupta 2020). Thus, women become victims of self-blame. For instance, in a study of research labs (Gupta 2020), when asked about the reason for gender inequality in the scientific organizations, the answers included: 'it's [gender inequality] the structure of the society itself'; 'gender equality means when both men and women use their full potential; ... no, it [gender equality] does not exist [in this institute] because women are not stretching their full potential ... they do not give their 100% [due to family roles]. It's their decision and you cannot blame others'; or, that equal opportunities exist and 'it is up to the women to avail those opportunities' (Gupta 2020: 697).

In viewing women as responsible for their lower status or women blaming themselves, there is guilt and humiliation for women that lead to a culture of silencing women's experiences due to further stigma of blame (Gupta et al. 2004).

'Triple stressors'

The un-level playing field created by the informal environment in the practice of science induces vulnerabilities. The male advantage, female disadvantage, and particularism in judgement of merit in science are the 'triple stressors' on women scientists. The triple stressors lead to physical, mental, and psychological exhaustion, although physical exhaustion is greater. In attempting to manage the stress, women often feel that life has become mechanical (Gupta 2001). Coping strategies often include compromising with one's career by forgoing opportunities (such as fellowships abroad), postponing research (women often become prolific researchers in their 50s), and redefining 'success' for a woman as one who can balance home and career (Gupta 2001). However, in attempting to balance both, often the career might be compromised.

However, gendered norms are not only reproduced at the scientific workplace, they are also challenged and reconstructed. Women scientists are not merely passive internalizers of social structures and meanings, but are also active interpretators of social realities and capable of modifying those. This occurs in various ways as discussed in Chapter 6.

Conclusions

To conclude, various norms and practices construct privileges and power for men in science organizations and consolidate institutions as the 'natural preserve of men' (Srivastava 2018: 42). For instance, interplay of hierarchy and lack of networking among women within their institute affects mentorship, visibility, and promotions. Hierarchical culture among the non-faculty staff and gendered assumptions in women's leadership abilities are some of the other aspects creating a masculine culture. Interpretation of incidents of discrimination is seen as unrelated to the formal system (equated with written rules), which keeps intact the 'rational' status of organizations.

The idea of women's primary domestic role gives rise to norms and practices that construct disadvantages for women in public spaces. For instance, the dual burden assumptions of male colleagues and biased treatment of pregnant employees. On one hand, work-life balance policies are inadequate; and on the other hand, such policies buttress a masculine environment if availed by women because men are either not the targets of those policies or they are not availed by men. Women's marginalization is best exemplified in the perception of women as 'privileged'.

The chief dilemma of women scientists is how to position themselves as 'scientists' and not as 'women scientists' in a contradictory context of a professed 'gender neutral' workplace alongside the existence of a gendered value system in reality. A masculine discourse, such as idealization of a scientist devoting long hours, and an assumption of women's inability to be similarly devoted, negatively impacts the perception of quality of women's research and the opportunities available to women. A woman's devotion to her family might be the expectation, but this expectation axiomatically categorizes her as not being committed to research. Some women might compromise career for family. However, such compromises imply structural constraints and a lack of social and institutional support.

The theory of critical mass assumes that an increase in proportions of minorities will positively impact their position within organizations. However, as this chapter shows, the problem is much more complex. An increase in proportions might diminish tokenism and the role of social stereotypes. However, the masculine culture, such as the practice of 'male model' of science, for instance, or belief in man as the 'ideal worker' and assumptions about the dual burden on women is unlikely to change with critical mass. The professed dual burden on women is a narrative that distracts from the workplace inequality. Hence, there is a need to go beyond the numbers. Gender in scientific institutions needs to be tackled to create a level-playing field for women. How much work has been done in this regard and what still needs to be done forms the subject of the next chapter. Finally, the latter also discusses women's agency in modifying norms and practices at the scientific workplace, thus exemplifying 'undoing' gender.

Notes

1 Narrated by Dr. Shubha Tole on Dec 12, 2018, in a InSDB—Indian Society for Developmental Biologists—session meeting at IITK (https://indiabioscience.org/events/insdb-biennial-meeting-2018)
2 IIT-Madras professor resigns, alleges caste discrimination, The Hindu, July 21, 2021: https://www.thehindu.com/news/national/tamil-nadu/iit-professor-resigns-alleges-caste-discrimination/article35089086.ece
3 The scientists in this study (Gupta 2017) were employed full-time and the nature of their work disallowed certain policies such as work-from-home. Nevertheless, these organizations consciously promoted worker-friendly culture through, for instance, family holidays, plush office space, high quality lunch provision, freely available tea and coffee throughout the day.
4 Business Standard. December 7, 2021. Available at: https://www.thehindubusinessline.com/economy/policy/govt-to-refund-employers-for-seven-weeks-of-maternity-leave-given-to-employees-wcd/article25506910.ece
5 The Gazette of India. Available at: https://labour.gov.in/sites/default/files/maternity_benefit_amendment_act2017_.pdf

References

Acker, Joan. (1990). Hierarchies, Jobs, Bodies: A Theory of Gendered Organizations. *Gender & Society* 4(2): 139–158.

Agashe, D., Maheshwary, S., Pattanaik JKumar, Prakash, J., Bhatt P., Arya, S.S., et al. (2022). Career Challenges for Young Independent Researchers in India. *Current Science* 122(2): 135–143.

Aiston, Sarah Jane, and Jung, Jisun. (2015). Women Academics and Research Productivity: An International Comparison. *Gender and Education* 27(3): 205–220. https://doi.org/10.1080/09540253.2015.1024617

Amin, A. (2021). *Calling Out Gender Bias In STEM. Rukhmabai Initiatives.* August 15, 2021. Available at: https://www.rukhmabai.com/calling-out-gender-bias-in-stem/. Accessed on September 6, 2024.

Ananth, Sudarshan. (2014). Women Leaders in Indian Science. *Current Science* 107(9): 1366.

Arora, C. (2022). Sustaining Science, Becoming Respectable. *Seminar: The Monthly Symposium* 760(December 2022: Women in Science): 26–32. Available at: https://www.india-seminar.com/semframe.html. Accessed on September 6, 2024.

Beck, U. (2000). *The Brave New World of Work*. Polity Press: Cambridge.

Belle, Deborah, Smith-Doerr, Laurel, and O'Brien, Lauren M. (2014). Gendered Networks: Professional Connections of Science and Engineering Faculty. In Vasilikie Demos, Catherine White Berheide, Marcia Texler Segal (eds.) *Gender Transformation in the Academy* (Advances in Gender Research, Volume 19), Emerald Group Publishing Limited, 153–175.

Belliappa, J. (2013). *Gender, Class and Reflexive Modernity in India*. Palgrave Macmillan: Basingstoke.

Benokraitis, Nijole V. (1998). Working in the Ivory Basement: Subtle Sex Discrimination in Higher Education. In Lynn H. Collins, Joan C. Chrisler and Kathryn Quina (eds), *Career Strategies for Women in Academe: Arming Athena*, Sage: New Delhi, 3–35.

Brumley, K.M. (2014). 'Now, We Have the Same Rights as Men to Keep Our Jobs': Gendered Perceptions of Opportunity and Obstacles in a Mexican Workplace. *Gender, Work & Organization* 21(3): 217–230.

Cabrera, S.F., and Thomas-Hunt, M.C. (2007). *"Street Cred" and the Executive Woman: The Effects of Gender Differences in Social Networks on Career Advancement* [Electronic version]. At https://ecommons.cornell.edu/items/eceb0818-fc03-4ba5-be45-fb1b0e4fda38. Accessed on September 6, 2024.

Cohen, L., Duberley, J., and Bustos Torres, B.A. (2023). Experiencing Gender Regimes: Accounts of Women Professors in Mexico, the UK and Sweden. *Work, Employment and Society* 37(2): 525–544.

Collinson, D., and Hearn, J. (1996). Breaking the Silence: On Men, Masculinities and Managements. In D.L. Collinson and J. Hearn (eds), *Men as Managers, Managers as Men,* Sage: London, 1–24.

Connell, R.W. (2000). *The Men and the Boys.* University of California Press: Berkeley.

Dogra, A., and Jayaraj, N. (2023). *Lab Hopping. A Journey to Find India's Women in Science.* Penguin: India.

Etzkowitz, H. (1971). The Male Sister: Sexual Separation of Labor in Society. *Journal of Marriage and Family* 33(3): 431–434.

Etzkowitz, H., Kemelgor, C., and Uzzi, B. (2000). *Athena Unbound: The Advancement of Women in Science and Technology.* Cambridge University Press: Cambridge.

Etzkowitz, H., Kemelgor, C., and Kehl, L.M. (2020). Gender Equity and Equality: Resistance and Advance in Academic Science and Innovation. In Helen Lawton Smith, Colette Henry, Henry Etzkowitz and Alexandra Poulovassilis (eds), *Gender, Science and Innovation*, Edward Elgar Publishers: Cambridge, MA, 380–404.

Fox, M.F., and Faver, C. (1985). Men, Women, And Publication Productivity Among Social Work Academics. *Sociological Quarterly* 26(4): 537–549.

GATI Report (2022). *Gender Advancement for Transforming Institutions. A Baseline Study.* February 2022. British Council and DST. Available at: https://gati.dst.gov.in/resources/wspublic;jsessionid=CE0B091C8F66AAC5B2C520A68D567495/resource-material/GATI-Report-February_2023.pdf. Accessed on September 6, 2024.

Grover, Shalini, Samantroy, Ellina, and Paiva, Nupur Dhingra. (2015). *"New India's" Inflexible Workforce – Caring Is But Women's Work'.* Guest Post. Available at: https://kafila.online/2015/05/31/new-indias-inflexible-workforce-caring-is-but-womens-work-shalini-grover-ellina-samantroy-nupur-dhingra-paiva/. Accessed on September 6, 2024.

Gupta, N. (2001). *Gender Inequality in the Work Environment at Institutes of Higher Learning in Science and Technology in India*, PhD dissertation, Indian Institute of Technology Kanpur. Available at: https://www.researchgate.net/publication/347885576_Thesis_Gupta. Accessed on September 6, 2024.

Gupta, N. (2007). Indian Women in Doctoral Education in Science and Engineering, A Study of Informal Milieu at The Reputed Indian Institutes of Technology. *Science Technology and Human Values* 32(5): 507–533.

Gupta, Namrata. (2010). Doctoral Research Environment in an Indian Institute of Higher Learning in Science and Technology. *Science Technology and Society* 15(1): 113–133.

Gupta, Namrata. (2013). *Women in Scientific Research in Government and Industrial Laboratories: A Comparative Study of Work Environment and Socio-Cultural Milieu.* Report submitted to Indian Council of Social Science Research, vide Fellowship no. F.No. 2–11/10/S.Fel. Available at: https://www.researchgate.net/profile/Namrata-Gupta-2/publications. Accessed on September 6, 2024.

Gupta, Namrata. (2016). Perceptions of the Work Environment: The Issue of Gender in Indian Scientific Research Institutes. *Indian Journal of Gender Studies* 23 (3): 437–466.

Gupta, Namrata. (2017). Gender Inequality in the Work Environment: A Study of Private Research Organizations in India. *Equality, Diversity and Inclusion: An International Journal* 36(3): 255–276.
Gupta, Namrata. (2020). Rationalizing Gender Inequality at Scientific Research Organizations, A Reproduction of The Indian Socio-Cultural Context. *Equality, Diversity and Inclusion, An International Journal* 39(6): 689–706.
Gupta, Namrata, Kemelgor, Carol, Fuchs, Stephan, and Etzkowitz, Henry. (2004). The 'Triple Burden': A Cross-Cultural Comparison of The Consequences of Discrimination. In Delilah Al-Khudhairy, Nicole Dewandre, Helen Wallace (eds), *Gender and Excellence in the Making*, European Commission: Brussels, 41–50. Available at: https://op.europa.eu/en/publication-detail/-/publication/ce10dfeb-4e36-49ed-a9ec-b2182c8b986f/language-en/format-PDF/source-288691746. Accessed on September 6, 2024.
Gupta, B.M., Kumar, Suresh, and Aggarwal, B.S. (1999). A Comparison of Productivity of Male and Female Scientists of CSIR. *Scientometrics* 45(2): 269–289.
Gupta, Namrata, and Sharma, Arun K. (2002). Women Academic Scientists in India. *Social Studies of Science* 32(6): 901–915.
Gupta, Namrata, and Sharma, Arun K. (2003). Gender Inequality in the Work Environment at Institutes of Higher Learning in Science and Technology in India. *Work, Employment and Society* 17(4): 597–616.
Hargens, L., McCann, J.C., and Reskin, B.F. (1978). Productivity and Reproductivity: Fertility and Professional Achievement among Research Scientists. *Social Forces* 57(September): 154–163.
Hasan, S.A., Sharma, Mahender Kumar, Khilnani, Sushila, and Luthra, R. (2012). Research Productivity of Female Research Scholars and Their Migration Pattern in Pursuit of Higher Education and Research. *Current Science* 103(6): 611.
Heuer, Mark. (2006). The Influence of Indian National Culture on Organizations. In Herbert J. Davis, Samir R. Chatterjee, and Mark Heuer (eds), *Management in India. Trends and Transition*, Response Books: New Delhi, 28–43.
Hofstede, G. (2001). *Culture's Consequences: Comparing Values, Behaviours, Institutions and Organizations Across Nations*. Sage: Thousand Oaks, CA.
Holt, H., and Lewis, S. (2009). 'You Can Stand on Your Head and Still End Up with Lower Pay': Gliding Segregation and Gendered Work Practices in Danish 'Family-friendly' Workplaces'. *Gender, Work and Organization* 18(s1): e202–e211.
Jayaraj, N. (2017): Women Scientists in India: What We Learnt from A Year of Lab-Hopping Across the Country. *Firstpost*, February 11, 2017. https://www.firstpost.com/living/women-scientists-in-india-what-we-learnt-from-a-year-of-lab-hopping-across-the-country-3274516.html. Accessed on September 6, 2024.
Joseph, M., and A. Robinson. (2014). 'Policy: Free Indian Science'. Nature 508 (7494): 36–38. Available at https://www.nature.com/news/policy-free-indian science-1.14956#/b2.
Kanter, R. (1977). *Men and Women of the Corporation*. Basic Books: New York.
Kawakami, C., White, J.B., and Langer, E.J. (2000). Mindful and Masculine: Freeing Women Leaders from the Constraints of Gender Roles. *Journal of Social Issues* 56(1): 49–63.
Kelan, E.K. (2010). Gender Logic and (Un)Doing Gender at Work. *Gender, Work and Organization* 17 (2): 174–194.
Krishna, V.V. (2001). Reflections on the Changing Status of Academic Science in India. *International Social Science Journal* 53(168): 231–246.
Kumar, N. (2009). Gendered Equation in Science: An Organizational Aberration? In Neelam Kumar (ed), *Women and Science in India,* Oxford University Press: New Delhi, 310–330.

Kurup, Anitha, Maithreyi, R., Kantharaju, B., and Godbole, R. (2010). Trained Scientific Women Power: How Much Are We Losing and Why? IAS–NIAS Research Report. Available at: http://eprints.nias.res.in/142/1/IAS-NIAS-Report.pdf. Accessed on September 6, 2024.

Levinson, D. (1978). *The Seasons of a Man's Life*. Knopf: New York.

Lewis, Suzan, and Humbert, Anne Laure. (2010) Discourse or Reality?: "Work-Life Balance", Flexible Working Policies and the Gendered Organization. *Equality, Diversity and Inclusion: An International Journal* 29(3): 239–254.

Lewis, Suzan, Gambles, Richenda, and Rapoport, Rhona. (2007). The Constraints of A 'Work–Life Balance' Approach: An International Perspective. *The International Journal of Human Resource Management* 18(3): 360–373. https://doi.org/10.1080/09585190601165577

Long, Scott J., and Fox, Mary F. (1995). Scientific Careers: Universalism and Particularism. *Annual Review of Sociology* 21(August): 45–71. https://doi.org/10.1146/annurev. so.21.080195.000401

Martin, Joanne. (1990). Deconstructing Organizational Taboos: The Suppression of Gender Conflict in Organizations. *Organization Science* 1(4): 339–359.

Masoodi, A. (2016). Where Are India's Female Scientists? *Livemint*. E-paper. April 19, 2016. Available at: https://www.livemint.com/Politics/N0j16kL2ZAzk94ssOKoTdK/Where-are-Indias-female-scientists.html. Accessed on September 6, 2024.

McPherson, J.M., Smith-Lovin, L., and Coo, J.M. (2001). Birds of a Feather: Homophily in Social Networks. *Annual Review of Sociology* 27(August): 415–444.

Morley, L., and Crossouard, B. (2015). *Women in Higher Education Leadership in South Asia: Rejection, Refusal, Reluctance, Revisioning*. Executive Summary. University of Sussex Centre for higher education and Research. British Council. Available at: https://www.britishcouncil.org/sites/default/files/women_in_higher_education_leadership_in_sa_-_executive_summary.pdf. Accessed on September 6, 2024.

Mukhopadhyay, C. C., and S. Seymour. (1994). 'Introduction and Theoretical Overview'. In Women, Education and Family Structure in India, edited by C. C. Mukhopadhyay and S. Seymour. 1–33. Boulder, CO: Westview Press.

Nature (2015). A Nation with Ambition. *Nature* 521(14 May 2015): 125. Available at: https://www.nature.com/news/a-nation-with-ambition-1.17520. Accessed on September 6, 2024.

Padavic, I., Ely, R., and Reid, E. (2019). Explaining the Persistence of Gender Inequality: The Work–family Narrative as a Social Defense against the 24/7 Work Culture. *Administrative Science Quarterly* 65(1): 61–111.

Palackal, A., Anderson, M., and Miller, B. Paige. (2007). Internet Equaliser? Gender Stratification and Normative Circumvention in Science. *Indian Journal of Gender Studies* 14(2): 231–257.

Paliwal, A. (2023). How India's Caste System Limits Diversity in Science — in Six Charts. *Nature* 613(7943): 230–234.

Pandey, Kartikay. (2007). Nose-Diving Scientific Temper: A Clarion Call for Science Enthusiasts. *Current Science* 92(1): 7.

Patel, R., and Parmentier, M.J.C. (2005). The Persistence of Traditional Gender Roles in the Information Technology Sector: A Study of Female Engineers in India. *Information Technologies and International Development* 2(3): 29–46.

Radhakrishnan, Smitha. (2011). *Appropriately Indian: Gender and Culture in a New Transnational Class*. Orient Blackswan: New Delhi.

Rajadhyaksha, Ujvala. (2012). Work-Life Balance in South East Asia: The Indian Experience. *South Asian Journal of Global Business Research* 1(1): 108–127.

Rajadhyaksha, U., and Velgach, S. (2015). What Is a Better Predictor of Work-Family Conflict in India? – Gender or Gender Role Ideology. In Liisa Makela and Vesaa Sutari (eds), *Work and Personal Life Interface in the International Career Context*, Springer: Berlin, 71–93.

Ramadoss, K., and Rajadhyaksha, U. (2012). Gender Differences in Commitment to Roles, Work– Family Conflict and Spousal Support. *Journal of Social Sciences* 33(2): 227–233.

Rao, S.S. (2013). Structural Exclusion in Everyday Institutional Life. Labelling Of Stigmatized Groups in an IIT. In G.B. Nambissan and S.S. Rao (eds), *Sociology of Education in India. Changing Contours and Emerging Concerns*, Oxford: New Delhi, 199–223.

Reskin, Barbara F. (1978). Sex Differentiation and the Social Organization of Science. In Jerry Gaston (ed), *The Sociology of Science*, Jossey-Bass: San Francisco, CA, 6–37.

Ridgeway, C.L. (2009). Framed Before We Know It: How Gender Shapes Social Relations. *Gender & Society* 23(2): 145–160.

Roth, W.D., and Sonnert, G. (2011). The Costs and Benefits of 'Red Tape': Antibureaucratic Structure and Gender Inequity in a Science Research Organization. *Social Studies of Science* 41(3): 385–409.

Sabharwal, Nidhi S., Henderson, Emily F., and Joseph, Roma Smart. (2020). Hidden Social Exclusion in Indian Academia: Gender, Caste and Conference Participation. *Gender and Education* 32(1): 27–42. https://doi.org/10.1080/09540253.2019.1685657

Sinha, J.B.P., and Kanungo, R.N. (1997). Context Sensitivity and Balancing in Indian Organizational Behaviour. *International Journal of Psychology* 32(2): 93–105.

Srivastava, Sanjay. (2018). Masculinity Studies and Feminism: Othering the Self, Engaging Theory. In Romit Chowdhury and Zaid Al Baset (eds), *Men and Feminism in India,* Routledge: South Asia Edition, 35–49.

Stack, S. (2004). Gender, Children and Research Productivity. *Research in Higher Education* 45(8): 891–920.

Subrahmanyan, Lalita. (2009). Women in Science in India. Has Feminism Passed them By? In Neelam Kumar (ed), *Women and Science in India,* OUP: New Delhi, 178–205.

Subramanian, J. (2007). Perceiving and Producing Merit, Gender and Doing Science in India. *Indian Journal of Gender Studies* 14(2): 259–284.

Thite, T. 2017. Women in Indian Science—A Young PhD's Perspective. Available at https://thelifeofscience.com/2017/09/11/women-indian-science young-phds-perspective/.

UNESCO (2022). *A Braided River: The Universe of Indian Women in Science*. By C. Coley, A. Dhillon, C. Gressel. UNESCO: New Delhi.

Wennerås, C., and Wold, A. (1997). Nepotism and Sexism in Peer-Review. *Nature* 387(6631): 341–343.

Williams, C.L. (1992). The Glass Escalator: Hidden Advantages for Men in the "Female" Professions. *Social Problems* 39(3): 253–267.

Xie, Y., and Shauman, K.A. (1998). Sex Differences in Research Productivity: New Evidence about an Old Puzzle. *American Sociological Review* 63(6): 847–870.

6

CHALLENGING GENDER INEQUALITY IN SCIENTIFIC RESEARCH

Introduction

Since gender (or, the behavior/roles attributed as masculine/feminine) is not biological but is socially constructed, challenging gender inequality would require destabilizing gendered norms, structures, and practices that privilege and impart power to men over women. This chapter analyzes how gender biases are being challenged at various levels, creating a potential for changing gendered norms and behaviors in STEM academia and research institutions. Women scientists' issues have increasingly gained prominence in the last two decades. A study (Malhotra 2018) of articles in *Current Science*, a peer-reviewed scientific journal published by the Indian Academy of Sciences, affirms the attention accorded to this issue. Women scientists have played a major role in fostering interest and awareness. This chapter analyzes the nature of the existing discourse, initiatives, and policies aiming at gender equity and ultimately gender equality.

Given the complexity of the issue, it is not possible either to cover all the problems or to suggest changes dealing with all of the facets of the problems. Nevertheless, the present understanding of the issue makes it possible to present a perspective on the policies and suggest the required changes therein. It indicates the structural changes that could help in moving towards gender equality and discusses the importance of men's participation in bringing about change. The chapter points out the need for future research to capture various aspects of the problems facing women in scientific research.

DOI: 10.4324/9781003596660-7

Does the pipeline thesis work?

The 'pipeline' model visualizes a single entry point and enhancing numbers in STEM education so as to maximize the numbers that reach in a linear path, the 'outlet' or a career. However, as shown earlier, a high proportion of women in doctoral education does not lead to a high proportion of women in research careers. For instance, despite about 38% of women obtaining doctorates (29% in engineering; 46% in science: AISHE 2021–2022, Table 36), there are only about 18.6% of women as researchers (S&T Indicators 2023: Table 22), lower than the world average of about 33.3% and even the average of Southeast Asia (26.3%) (UNESCO Science Report 2021). If the pipeline is visualized as different for various fields of STEM, some fields are stereotypically male or female at the entry point, but end up being male. For instance, higher proportions of women obtain doctorates in biosciences than physics. Yet proportions as faculty in prestigious institutes or in leadership roles remain low in both cases, as shown in Chapter 4.

A gendered accumulation of merit during the doctoral period, construction of male privilege, and female disadvantages at the STEM workplaces fail to translate degrees into careers, and once in a career, into high positions or awards. Instead of a linear 'pipeline' model, there is a need for a recognition of multiple entry and exit points in STEM workforce development (rather than 'leaks'), a variety of training paths converging into a career, a model conceptualized as a 'braided river' (Gardner-Vandy et al. 2022). Apart from this, one needs to take into account the social context that constructs fields as masculine and feminine, which renders difficult any single model of career development. One of the ways to create opportunities for women for a successful career is to enhance their access to quality education at all levels and in all STEM fields. A prominent effort to improve such access has been the introduction of affirmative action in the IITs.

Affirmative action in the IITs

Due to the declining numbers of women admitted to the IITs and a gender imbalance, a committee headed by the director of IIT Mandi, Professor Gonsalves, was set up in 2016 by the Joint Admission Board (JAB) of the IITs to make recommendations for increasing female enrollment in the IITs. This report confirmed the role of biases that limit access to coaching and the gendered view of branches and proximity to home that determine choices for women during counseling (JAB 2017). The committee found that 1,400 women within the range of ranks that received admission did not join the IITs, and another 2,400 women qualified JEE (Advanced) but with lower ranks. Hence, this committee recommended supernumerary seats for women in its report in 2017. Following these recommendations, the central government decided to raise the number of women enrolled in undergraduate courses to 20% by 2020 in these institutes, with the intention to phase out the supernumerary seats by 2026. The proportion has been achieved with the percentage of female candidates admitted to IITs in 2024 at 19.75% (JEE Report 2024). Since only those who qualified JEE (Mains) and then JEE (Advanced) make it to the new merit list, merit was not compromised in any major way. The cutoff for JEE (Advanced) is irrespective of gender. Among the long-term measures, the report recommended efforts to tap the potential of girls from Class 8 and counseling for parents, teachers, and girls for encouraging girls to aim for top institutes.

The affirmative action policy to increase the proportion of girls is helpful from the point of view of utilizing human resources and of gender equity. As seen in Chapter 2, a lack of critical mass fosters and reproduces masculinity in the IITs. An increase in number of women in the IITs will help in creating a healthier environment. This might be evident already as noted in the GATI (Gender Advancement for Transforming Institutions) Report (2022). Nevertheless, there are also reports of the formula being seen as a reservation quota and a privilege for women by the male students (Srivastava 2023). This suggests the need to educate men about the necessity for quotas for women and to undertake measures suggested by the committee so as to phase out the quotas in the near future.

Gender equity measures are being taken by the institutes themselves, such as some IITs providing career counseling and support for women graduates. For example, IIT Madras's alumni-supported initiative to guide and support women graduates through professional networking, scholarships, and mentorships or NIT Jaipur reserving 30% scholarships for women (UNESCO 2022). Tata Institute of Fundamental Research's Vigyan Vidushi program for women halfway into master's courses (in physics, math, and computer and system science), aims to expose students to advanced topics in research and to be mentored by successful women scientists in order to 'counterbalance the conflicting messages' imposed on girls by Indian culture.[1] However, there is a need to study the impact of such measures.

Women scientists' struggle against gender inequality in STEM research

Gendered norms are not only reproduced at the workplace, they are also challenged and reconstructed. The challenge by women scientists is visible and has gathered pace in the last two decades. In the first half of the twentieth century, speaking out against discrimination was unthinkable as the mere survival of women in science was a challenge (Sur 2009). In the 1990s, Subrahmanyan (1998), while researching women in academic science, felt that 'feminism has indeed passed these women by' (p. 281); that while the women scientists were not totally non-feminists, without their involvement in the overall women's movement, their problems could not be solved.

However, women scientists are now challenging gender discrimination in various ways. Women scientists have taken various initiatives for mentorship, conducting lectures/workshops on the topic, and surveys.[2] They give expert advice to the government through membership of the Inter-Academy Panel of IASc-INSA-NASI. They have taken the lead in visiblizing the marginalization of women in science through participation in panels, and creating exclusive website and internet resources. Such websites have been formed under the aegis of the Indian Academy of Sciences (IASc) and the Inter-Academy panel.[3] The global forums of women scientists (e.g., TWAS, AWIS) and subject-specific conferences and groups, such as International Conference on Women in Physics, have provided platforms for women in science in India to voice their issues and understand the country-specific as well as global parameters of their problems. The discussions have become sharper in attacking the science institutions as gender biased in recent years, for instance, a group of women physicists is looking at sexism and patriarchy in their discipline (Priyadarshini 2023). There might be a cost, though, to women scientists actively working for gender equality in science as their role as activists might get greater prominence than their work (Dogra and Jayaraj 2023).

Nevertheless, the coming together of women scientists has created a discourse within the establishment on gender in scientific research careers, particularly after the NIAS report on women in science in 2004, which highlighted the low proportions of women scientists in elite institutes and fewer reward recipients among women. The efforts of women scientists, combined with the government willingness to find solutions to the various problems facing women scientists, have led to the survey-based reports sponsored by the government and various science academies (e.g., INSA 2004; DST 2010; Niti Ayog 2017). Further, the narratives of women scientists brought together by the social scientists (e.g., EPW 2017), and by the women scientists themselves (e.g. Lilavati's Daughters), the media's interviews with women scientists at The Life of Science platform since 2016, and collaboration between social scientists and scientists have refined the understanding on the issue (e.g., Kurup et al. 2010). The discourse on gender in science has progressed considerably in the last 20 years from highlighting a lack of female participation to gender inequalities within the scientific institutes and structures.

At the micro level, women scientists are actively engaged in transforming their position in science. There are instances of gender rebellion by women scientists. For instance, the 'MeToo' movement called out sexual harassers in academia. Two women scientists founded a media platform, 'BiasWatchIndia', that highlights women's underrepresentation and gender-biased panels in Indian science conferences, meetings, etc. Mayurika, a cancer biologist at IISER, Pune, started a daycare revolution at her institute.

Beginning with a request for daycare for her daughter and subsequent allocation of a room by the director, she went on to set up a full-scale daycare center (Dogra 2016). Some other instances include the physicists who disproved the theory that competent women physicists are too few to bring out a journal issue by themselves, when ultimately three scientific journals brought out special issues by women physicists (Chowdhury 2017); the rebellion of a Dalit female doctoral student, studying nanomedicine, who won the legal battle against casteism in her institute (Chander 2021); and, a female director of an institute sacking an award winning scientist for sexual harassment (Jayaraj 2021). Such fights, although small in nature, actively break stereotypes, inspire others and contribute to challenging gender inequality in science.

Women scientists play an important role through mentorship. Senior women scientists' mentorship could be very significant particularly for young women researchers. As Dogra and Jayaraj (2023) found that parents allow daughters to leave home to pursue research if a guide is female, and a woman scientist can impart confidence to rebel against an arranged marriage to follow their passion for science (p. 76–78). Women scientists have created systematic mentoring initiatives, as for instance, Indian Women and Mathematics (IWM), which aims to raise awareness about current research in mathematics and provides opportunities to graduates and early researchers to meet other women mathematicians. According to Dr Kang, a well-known Indian woman scientist, mentorship could be the most valuable approach to bridging the gender gap, and senior women that have gone through similar experiences are best positioned to advise personally and professionally (Dogra and Jayaraj 2023: 47).

The efforts of women scientists to promote gender equality have provided the much-needed impetus and information for government policy.

Government and the changing discourse

The significance of the role of women scientists in Indian science and innovation is being increasingly recognized by the government. It was witnessed first in the 2013 Science, Technology and Innovation Policy (STIP), which professed 'gender parity' as one of its aims. Schemes to encourage women scientists to re-enter career after a break have been existent since 2002–2003 and now restructured under Knowledge Involvement in Research Advancement through Nurturing (KIRAN). However, the government took a leap forward in 2020 with the Draft STIP and the GATI Initiative. Schemes under KIRAN have been revised recently to include, for instance, postdoctoral fellowships (WISE-PDF) and fellowships for projects that benefit society (DST-SCOPE). Before discussing the change, it is essential to briefly note the nature of policies.

Various government initiatives that began in the first decade of the millennium include the following strands: a, *those that target women as individuals*, such as fellowships for women to bring back women scientists after career breaks through extramural projects, train women in Intellectual Property Rights, encourage entrepreneurship, schemes for women scientists to collaborate abroad, awards for women scientists, portals for women in STEM, and postdoctoral fellowships; b, *specific schemes to enable women scientists to achieve work-life balance,* such as the Mobility Scheme. Also, the women scientists in the government sector are beneficiaries of the generous government rules for paid maternity leave (26 weeks for two surviving children and 12 weeks thereafter) and childcare leave to women and (since 2018) single male employees having minor children below the age of 18 years, for a maximum period of two years (i.e., 730 days) during their entire service, for taking care of up to two children; c, *schemes to enhance women's engagement with education in STEM* (e.g., Inspire, Vigyanjyoti); d, *female-only institutes as targets*: upgrading women's universities through CURIE; and e, *targeting institutions* (GATI).

Apart from this, government has also appointed a Standing Committee for Promoting Women in Science in 2016;[4] to hold workshops, conferences, etc. However, some of the problems related to the policies, although unique and well-intentioned, are discussed in the following sections.

The implementation hiatus

There is a gap between the policies and their implementation. For instance, a woman scientist, designated as principal investigator (PI) under the Women Scientist Scheme-A (WOS-A), might face harassment from the staff of the government scientific agencies for the release of funds for their project (Gangopadhyay 2017). The claim that 30% of women got permanent jobs in academia due to the WOS-A Scheme appears unsubstantiated by data (Gangopadhyay and Das 2016).

The gendered philosophy of 'separation of spheres', where women's primary role is domestic, that the woman is the center of the family and is supposed to prioritize family over career, is reproduced through many policies. For example, a publication of DST-Centre for Policy Research (p.13) states:

> WOS-C Scheme aims to train women having qualifications in science/engineering/medicine or allied areas in the area of IPR and their management for a period of one year. The training will allow them to work from their homes *and thus maintain a good balance between professional and domestic demands.*
>
> *(Tewari and Bhardwaj 2018: 13)*

Similarly, the Mobility Scheme is aimed to provide an opportunity to women scientists who are facing difficulties in their present job due to relocation (owing to marriage, transfer of husband to any other location within the country, attending ailing parents, and accompanying children studying in a different city), and is expected to act as a filler while searching for other career options at the new place. The initiative intends to provide a harmonious environment during early phases of women scientists' careers *where they would like to stay active in research in addition to attending and fulfilling other responsibilities in the domestic front.*[5]

Women scientists recognize the gendered nature of the policies aimed for their benefit. For instance, Godbole (2017), a physicist at IISR, Bangalore, while acknowledging that the Women Scientist Scheme of DST has led to an increase in the proportion of women principal investigators, remarked on the Mobility Scheme, 'We have programmes to come back from a break but not for minimizing/avoiding the breaks and their after effects ... *Such schemes should be meant for couples, with either of the partners doing the relocation or the redefinition*' (Godbole 2017: 672). According to Vaidya (2017: 83), a neuroscientist at TIFR:

> When you single out women as the sole caregivers, it fails to achieve the broader change that is needed to level the playing field. However, if maternity–paternity leave policies were discussed collectively as the best way to ensure that families can generate the required time to manage child-rearing and their work, it would send a different message.

That the childcare leave should be extended to men, or equally divided between the parents, has been recommended by women scientists and even the courts.[6] In one of the few gender-neutral measures, crèche facilities have been provided by the Maternity Benefit Act 2017 for organizations with 50 or more employees; however, creche guidelines framed in 2018 have been legislated by only a few states, and how successfully they are implemented by the employers has yet to be seen.[7]

Since 2005, the Government of India has adopted 'gender budgeting' at the levels of the planning, budget formulation, and implementation processes. It is seen as a major instrument of gender mainstreaming. However, there are several flaws in the Gender Budgeting Statement (GBS), such as errors in the process used for determining and reporting the estimates. The preparation of the GBS for the union budget largely remains an ex-post exercise, with most of the ministries/departments providing information on their schemes for reporting in the GBS after allocations for schemes have been made (Mehta 2020). Also, there remains a big gap between budgetary outlay and outcome due to non-utilization of the allocated funds. Gender commitments do not get translated into budgetary commitments in the absence of funds, functions, and functionaries (Rudra 2018). The concept does not entail mere increased outlays for the women-centric programmes; rather, it demands financially effective decisions of the public sector based on the principles of gender equality and analysis of gender-differentiated impact of the earmarked public expenditure allocations (Rudra 2018).

At times, the discourse on gender parity among scientists and administrators has resulted in measures that further segregate women and fail to bring the intended respect to them in the scientific community. For instance, the Women Scientists' Awards tend to downgrade women by excluding them from the 'gender neutral' awards. Similarly, Women's Science Congress (WSC) segregates women from the traditional Indian Science Congress (ISC), although the latter is increasingly losing its sheen (Sinha 2023) and is mired in controversies leading to postponement of ISC 2024. Women scientists have questioned the rationale for holding a separate WSC and have called for scrapping it (Ashok 2018). In a striking change, the 108th ISC in 2023 focused on women's empowerment, although concerns were raised over the commitment to the theme; and the WSC in 2023, held as part of ISC, failed to generate any interest, with more than three-fourths seats remaining empty on the first day itself.[8]

Separate awards and conferences segregate the scientists as 'men' and 'women', thus reproducing gender hierarchy in science. For women scientists to be recognized as 'scientists' and not as 'women scientists' requires their integration with the mainstream scientific community.

The changing discourse

Since the end of the second decade of this millennium, there are indications of a change in discourse from the mere emphasis on 'gender parity' to 'gender equity' in science, from enhancing participation to creating an enabling environment and equal opportunities. The draft Science, Technology and Innovation Policy (STIP) 2020 stresses on 'equity' and 'inclusiveness', aims to mandate the representation of women in all scientific decision-making bodies to one-third, envisions the inclusion of LGBTQ+ community in gender equity conversations, focuses on the need for sensitization, gender non-specific policies, sex-disaggregated data, and a deeper understanding of equity and inclusion issues. Age criteria for awards and fellowships have been expanded. For instance, more awards have been introduced, such as Vigyan Ratna and Vigyan Shastri, that have no age limits; WISE-SCOPE and WISE-PDF fellowships have age limits of 60 years.

An increasing recognition of the implicit biases in the system among the policymakers is indicated in the GATI (Gender Advancement for Transforming Institutions) Charter as part of the GATI program launched in 2020 by DST.[9] The program draws essence from Athena SWAN in the UK and in partnership with such accredited UK universities, targets institutions nudging them to create an *enabling environment* (GATI Charter) for women in STEM at all levels, throughout their career. It is premised on the belief that 'the quality of an institution depends on its capacity to create *opportunities* ... for growth of all' (Jolly 2022: 35) and recognizes the goal of *'gender equality'*.

The words in italics indicate the change in vocabulary, from parity to equality. GATI is the first initiative of the government that targets organizations, particularly the reputed institutes and not individual women and focuses not merely on enhancing numbers, but on changing organizational cultures and systems. Significantly, it makes the institutes admit gender inequality through its charter, recognizes various parameters of inequality, and aims to create awareness through self-assessment, prodding them to adopt gender equality measures through its accreditation system. The process involves the much-needed collection of quantitative and qualitative data to understand gender inequality. The collection of quantitative data covers gender disaggregated data encompassing all stages of a career and accounts for the transgender.

Some positive impacts of GATI are already being observed. IIT Delhi, for instance, has opened an office of diversity and inclusion; some of the premier institutes, such as IIT Kanpur, are conducting special drives for the recruitment of women faculty; and IIT Bombay has started a special course on gender sensitization for students. Initiatives, such as proactively nominating women for external awards, facilitating women as principal investigators, ensuring at least one woman as part of decision-making committees, are being undertaken by various institutes. Institutes are looking at the issues of campus safety and discrimination in hostel rules and enhancing efficacy of internal complaints cells on sexual harassment. Plans are afoot for better infrastructure for work-life balance, and discussions are being held on flexible timings and meeting schedulings for non-residential campuses (Jolly 2022).

The possible weaknesses of GATI

The project is at its nascent stage as yet. Although noble and comprehensive in its conceptualization, the program could have its pitfalls in implementation. The measures are likely to encounter traditional norms of patriarchy and 'merit'. For instance, while special recruitment drives for women are being undertaken, it is not clear if they would hire women with meandered careers (women with breaks or those working on research projects). If we were to 'normalize the idea of individuals changing pace and direction' (Batchelor et al. 2021) and recognize that careers need not move in a linear fashion, it would open the doors for acceptance of 'break', leave, and late careers.

The fundamental element in achieving the project's aims is to convince the scientific community that there is gender inequality. Jolly, the PI of GATI, notes the disproportionate burden that the GATI implementation puts on women; the need for a credible community engagement and support of men within the stakeholder community (Jolly 2022). However, how we convince the scientific community is a major challenge. This requires the collection of institute-wise data on the participation of men and women scientists; their career growth over the years, and qualitative data on perceptions, conducting gender awareness programs, and research studies on gender (as discussed later). Much of this is being covered under a 'GATI self-assessment application' and data templates. However, this data needs to be made public and accessible to all. Its compilation over a few years will help to understand the prevalent situation and the progress of GATI.

The GATI program, with its focus on promoting women in STEM institutes, appears to be a measure for the benefit of women. It might be important to highlight that the measures to eliminate gender inequality are likely to improve the work climate for men also. Further, can we convince the faculty members in such institutes that equity measures, such as the special recruitment drives, will not compromise on the quality of the faculty or the so-called 'merit'? Can we counter the perception that women get things easily, that men are the victims of the gender discourse, while women are the beneficiaries? Research shows that

> *... progress toward gender equality will be slowed to the extent that efforts are focused exclusively on women ... that to be effective, expanding efforts to include men requires a broad vision.*
>
> *(Padavic et al. 2019: 43).*

Fundamental changes towards gender equality require organizational reforms that might not be women-specific. Such reforms include transparency, addressing issues of hierarchical culture, and masculine culture in day-to-day interactions. It is difficult to see how this could be achieved with GATI. Further, departments within an institute are responsible for recruitment, promotion, providing research conditions, etc. Given the fact that the departmental cultures vary, efficacy of measures would require department-specific initiatives. Will the institutes implementing GATI be able to account for such complexities?

Further, will the implementation of GATI lead to 'competing inequalities' (Bhopal and Henderson 2019)? A study of Athena SWAN and race equality projects found that there has been a privileging of gender over race in terms of addressing inequalities in higher education (Bhopal and Henderson 2019).

In India, although GATI self-assessment framework does emphasize intersectionality, would its implementation lead to middle-class and upper-caste women as primary beneficiaries of the program and sharpen the inequalities within women in science?

Finally, while surveys are recognized as an important mechanism for feedback in GATI, comprehensive feedback requires not only to capture the opinions and experiences of women, but also of men, to understand how the community as a whole is responding to GATI. Further, as Schmidt et al. (2020) suggest, that to manage Athena SWAN's dynamic complexities, unintended consequences, and troublesome incentives, developing 'highly qualified local implementation professionals' is essential. Perhaps, this aspect might be looked into further.

Studies on the impact of Athena SWAN (on which, GATI is modeled), show that while there has been a positive impact on women's careers (ECU 2014), there are resentments about perceived positive discrimination, perceptions of the continuation of entrenched power imbalances, work-life imbalance, questions about the sustainability of positive changes, and the belief that achieving the award could become an end in itself (Ovseiko et al. 2017). Perhaps, knowledge of such pitfalls might help us to prevent those.

Nevertheless, the GATI program is significant as it represents an effort to view gender inequality as part of systemic inequality, and not only as an issue of a skewed ratio or women's work-life balance problem. It forces the institutes to acknowledge their responsibility towards promoting gender equality. More importantly, it encourages discourse on gender to recognize the problems and think of solutions within the institutes/labs and universities.

Equity or equality: conceptualizing change

Socio-cultural norms construct inequalities based on biological differences. Hence, enabling a safe environment for women requires overcoming the specific obstacles that women face so as to enable a level-playing field. These call for implementing measures of 'gender equity'. Gender equity means fairness of treatment for men and women according to their respective needs. However, as the European Institute for Gender Equality notes in its definition of 'gender equity':[10]

"The term 'gender equity' has sometimes been used in a way that perpetuates stereotypes about women's roles in society, suggesting that women should be treated 'fairly' in accordance with the roles that they carry out. This understanding risks perpetuating unequal gender relations and solidifying gender stereotypes that are detrimental to women. Therefore, the term should be used with caution to ensure it is not masking a reluctance to speak more openly about discrimination and inequality".

Focusing on 'difference' tends to reinforce stereotypes. Gender equity cannot be a goal by itself. The goal is to achieve 'gender equality'. Gender equality does not merely mean equality in numbers. It is not the same as gender parity. 'Gender parity' measures gender balance in a given situation and concerns 'relative equality in terms of numbers and proportions of women and men, girls, and boys, and is often calculated as the ratio of female-to-male values for a given indicator'.[11] Gender equality, as defined by the European Union, does not mean that women and men will become the same, but that women's and men's rights, responsibilities, and opportunities will not depend on whether they are born female or male. Gender equality is not a women's issue but should concern and fully engage men as well as women. Equality between women and men is seen both as a human rights issue and as a precondition for, and indicator of, sustainable, people-centered development.[12]

The obstacles for women scientists in the Indian context arise from a rigid dichotomy of public-private spheres, from a historic growth of science and its institutions, and India's place in a developing world economy and research network, which have created science institutions and organizations as male-dominated and masculine. Hence, there is an urgent need for gender equity measures, but we also need to move forward towards gender equality targeting the structures and processes that create gender inequality.

Government policies help in addressing the issues of gender equity. However, more such initiatives need to be thought out. Ageism is a pernicious issue raised time and again, and more recently pointed out in the Draft STIP 2020 (7.2.3 of STIP 2020) of the government. Dealing with ageism, for instance, requires relaxation of age for recruitment and provisions for lateral recruitment. Gender equity for a level-playing field calls for greater sensitivity to women's issues and recognizing biases.

However, steps to create an enabling environment for women are unlikely to yield results unless systems generating inequality are targeted and gender equality is the goal. The difference between gender equity and gender equality might be demonstrated through an example of networking. Creating women's networks is a gender equity measure, which is immensely useful in creating mentorship and visibility. However, they have limitations. Such networks are unable to deal with the exclusion of women from powerful networks, which are usually male (Carbera et al. 2007). Gender-neutral networks could bring more equality between men and women in research (Checchi et al. 2015). The way forward that aims for gender equality would conceptualize making informal interactions and networking between the scientists within and outside an institute easier for both men and women. This and other areas of concern in moving towards gender equality are dealt with in the next section.

Towards 'gender equality': chief areas of concern

Moving towards gender equality *in scientific research* requires measures of gender equity that take into account the specific needs of women, combined with those leading to gender equality and an integration of women scientists within the scientific community. Chief aspects that need consideration are: a, interaction issues; b, organizational reforms; c, reforming institutions of science; d, work-life balance structures; e, dealing with biological differences; and f, integrating caste and gender.

Structures of interaction

Interaction could become a site for 'undoing gender', or dismantling gender inequality (Deutsch 2007). Interaction between scientists supports exchange of ideas, resources, networking and collaborations. However, as discussed in the earlier chapters, women lack such interaction. Hence, structural changes that improve informal interaction could change the work climate. A combination of decentralized day-to-day working, setting up weekly informal meetings, and creating architectural changes could be helpful in increasing informal interaction within the institutes.

In a study on CSIR labs (Gupta 2016), one example of a possible model of change was found in a center for interdisciplinary research that was set up in a CSIR lab in 2010. The building which houses the center incorporates the construction style which is different from the traditional. Instead of offices arranged in a long dark corridor, it has a circular design which is more interaction-friendly. The members of the center have established a coffee club which meets once a week. The work environment here was perceived as quite positive and had an impact on gender relations. Significantly, this center has a decentralized system of management. It does not have an overarching HOD and is headed by a collegium of five scientists, including two women scientists at the time of the study, who hold office for two years. The positive environment here was attributed by the scientists in the Gupta (2016) study, to lesser hierarchy and greater space for informal interactions. For instance, Anamika (female, Scientist E1), when asked how she perceives gender equality at her workplace, said (Gupta 2016: 455):

> Gender equality is [achieved] when people surrounding you perceive you as a colleague and not a 'woman colleague' ... Yes, it [gender equality] exists in my group. It's a collaborative and cooperative environment. People talk about their research, seniors are trying to be your mentor [she just joined about a year ago]; there is a Friday coffee club where I go, we discuss, debate. There is comfort in the work environment'.

This study found that the centre had considerable internal collaboration and mutual interaction and had a robust output in terms of research productivity. However, it has a miniscule impact on the entire institute, and such centres are rare. Their positive role needs to be highlighted so that similar models can be established in other institutes as well. Apart from creating network of connections that are gender-neutral, an inclusive informal interaction helps in countering competence bias against women (Ridgeway and Correll 2000), and could be useful in moving up the academic career ladder.

Another initiative encouraging interaction existing in many parts of the world is that of 'mentoring'. Mentoring programs (MP) are built on the idea of older, senior professionals sharing their experiences and networks with younger professionals in order to help them advance in their careers (Powell 2018). In India, MP usually implies interaction of the young women in science with the senior and established women professionals through workshops and seminars. The Department of Science and Technology (DST) and the Science Academies conduct mentorship and awareness workshops, addressing the student community, executives in industry or administrators in high level institutions, etc. The University Grants Commission held a Capacity Building of Women Managers in Higher Education (CBWM) program between 2003–2013 that included sensitization, awareness, and motivation workshops for women academics in universities.

Mentoring initiatives *for students* aspiring for a career in scientific research and for tapping the potential in smaller cities are extremely helpful in creating information about the opportunities and the resources available. However, mentorship programs *for sensitizing or motivating the faculty* are more about 'fixing the women' rather than 'fixing the system'.

It might be more useful to design formal mentorship programs within the academic/research system that aims at the mentoring of the newly recruited women (and men) scientists so as to enable them to use their potential fully and to rise in organizational hierarchy. For instance, in many US universities, new faculty members are assigned to those ahead of them (Etzkowitz et al. 2000). These seniors act as mentors, introducing them to other faculty members, inviting them in social functions, and giving advice, tips, and strategies which help in knowing policies and also create a sense of what is required/expected of them. Compared to informal mentorship, formal mentoring programs are a more transparent, formalized, and egalitarian way of providing mentors (Powell 2018).

Organizational reforms

Invisible barriers operate in organizations that mostly arise from the stereotypes about female competency as a scientist and a leader. Of the various measures, three are most pertinent to counter these beliefs. They are affirmative action, organizational efficiency, and transparency.

Affirmative action

Affirmative action in recruitment has been suggested as a solution to underrepresentation worldwide. Affirmative action, including quotas, has been justified on the grounds of historical exclusion and contemporary biases against the disadvantaged groups that create threats to fair treatment (Ceci and Williams 2015). In India, a quota system has usually been rejected by women scientists. Reviewing *Current Science* articles on quotas for women scientists, Malhotra (2018) found in the 1990s that 'The quota system is seen as not being in the interest of women as it could be perceived as "patronizing" and could also lead to a "feeling of disgust" amongst women and "resentment" amongst men' (p. 1717).

However, affirmative action does not necessarily imply quota. In the US and many developed countries, such as DACH countries (Germany, Austria and Switzerland), it implies gender-based preferential selections when job candidates are equally competent. Affirmative action programs have been followed in Sweden in case of students and professorships and in the EU Horizon programs for research funding. In India, adoption of the GATI Charter has prompted special recruitment drives for women faculty. The nature and outcomes of such drives and how they are being perceived by the scientific community is not clear yet.

Studies show that affirmative action policies may successfully attract female job seekers (Ibanez and Riener 2018); although, some studies have reported negative expectations about the competence of those hired (Heilman and Alcott, 2001). Further, men who are the non-beneficiaries may feel discriminated against (Heilman 1996). However, recent studies show more positive outcomes. For instance, Henningsen et al. (2022) found that female and male evaluators were more likely to rank the female applicant as their first choice across selection policy conditions. These results, according to the authors, support the recent claims that scientists seem to have internalized society's goal to increase women's representation in science. Affirmative action has helped to produce modification of the presumptions of competence for women as compared to men (Ridgeway and Correll 2000).

Gender biases persist not just at the entry level but at all stages of a career. Fixing the percentage of women for various selection/promotion committees and at the policy and planning stages has been suggested due to the argument that women and men perceive women's problems differently (Kurup and Maithreyi 2011; Sinha and Sinha 2011). The assumption behind such a quota is that men tend to select men. Research studies on quota in evaluation committees show that the success of such measures is mixed. Gender quotas have been introduced in selection boards of European countries such as

Norway, Sweden, Finland, and Spain where the compulsory share for both genders is 40%; this has resulted in limited success. Gender quotas in evaluation committees are not sufficient to raise the representation of women, nevertheless presence of women makes committee members more aware of gender issues (Wallon et al. 2015).

Organizational efficiency

Gendered practices, particularly, work practices that are constrained by gendered images of competence and organized around gendered ideas of who is an 'ideal worker', can be inefficient organizational practices; such practices are wasteful and are unproductive for both men and women (Rapoport et al. 2002). These include, for instance, idealizing the devotion of long hours to the lab/workplace but unrelated to the quantity and quality of the work performed; meetings without much agenda or urgency; or bureaucratic red tape at lower levels. There is a need to highlight such wasteful practices.

Efficiency of women's work is affected due to biases in the allotment of resources. Problems, such as women academics lacking research scholars or resources, are reflective of deep-rooted biases. A gender audit of departments will help to unravel the discrepancies in resources, which is the first step towards addressing any discrepancy. Rapoport et al. (2002) suggest a practical approach to bring about gender equality referred to as 'dual agenda'. This approach assumes that by making changes in work practices to enhance gender equality, it is possible to enhance workplace performance and organizational effectiveness (Charlesworth and Baird 2007).

Transparency

As noted in the previous chapter, committees for recruitments and promotions are known to ask gender-based questions. However, the system is not transparent; hence, it is difficult to say that there is gender-bias in actual selections. Studies worldwide have shown that gender bias is more likely to occur when assessments are based on obscure criteria and the process of evaluation is kept confidential (e.g., Husu 2000; Rees 2004). Researchers have advocated transparency and accountability as the chief instruments of gender equality. Transparency in the system and accountability for evaluations is important, because '... when evaluators are accountable for their assessments — in that they must be prepared to explain them to others — they look more closely at the information available to them and consider the impression their decisions will make. Both greater cognitive scrutiny and concern for appearances reduce the extent to which people rely on implicit stereotypes to form their judgments of others' (Ridgeway and Correll 2000: 117).

Gender biases in recruitment can occur even before applicants are evaluated (Nielsen 2016). To prevent gender biases at the key decision-making points, suggestions have been made to hold evidence-based workshops and trainings involving the decision-makers, such as hiring committees, promotion and tenure committees, and blinding of demographic characteristics (such as gender) to rate the applicants (Moss-Racusin et al. 2021).

To ensure transparency in appointments, apart from open advertisement and clear criteria, many universities elsewhere require setting out proceedings of the process and decisions before the university boards and the presence of HRM advisor to professionalize the appointment process (van den Brink, Benschop and Jansen 2010). However, transparency and accountability are difficult to achieve. Micropolitics is known to obstruct the implementation of transparency protocols (van de Brink, Benschop and Jensen 2010). Institutions can be required to be more explicit and transparent about how merit is evaluated while incentivizing the departments with funding allocations (Llorens et al. 2021). Transparency protocols and the process and decisions being made visible for the larger academic society already exist in Sweden and Finland.

Reforming institutions of science in India

There are other grounds for inequality apart from discriminatory structures in organizations. These include inequalities in institutions of science (grant agencies, award committees, and scientific academies) and the doctoral/postdoctoral education system.

Reforming the institutions of science implies creating an educational and research ecosystem that provides equal opportunities for all desirous of a career in S&T. This requires, first and foremost, strengthening of our universities and raising the quality of doctoral education. This necessitates greater research funding and structural changes in universities to encourage research. There is a need to think about greater encouragement to women in doctoral and postdoctoral research, while strengthening grievance redressal mechanisms against intimidation and harassment (of men and women) during the program. Women (and men) researchers would benefit from opportunities generated through greater collaboration between academic institutes and between academia and industry. Based on the analysis in the earlier chapters, an area for reform includes streamlining the methods for securing external funding. Systems of awards and recognition such as fellowship of academies that require nomination by existing fellows and impart power to the few who decide, could be reworked to include a practice of self-nomination and power to more than one independent committee (Gupta 2007). The recently instituted Rashtriya Vigyan Puraskar accepts self-nominations (RVP 2025). More mechanisms might be thought of so that the recipients of awards represent a diverse community of scientists.

There is also a need for a data collection agency. Lack of data is a roadblock in understanding gender issues and thereby tackling them. While we have enrollment figures for women in various degrees (UG, PG, and PhD), we do not know how many women (and men) drop out at the national level. Hence, it is imperative to have data on this. The initiative of collecting data by the Ministry of Education through the NIRF (National Institutional Ranking Framework) is very helpful for researchers in social sciences. However, the NIRF data from 2019 onwards has actually omitted the information regarding the gender-wise breakup of faculty strength. The recent GATI Report (2022) provides the much needed data based on a sample of institutes, but more extensive surveys covering larger numbers of institutes are required to make sense of women's positions in all types of institutes and at various levels of one's career.

This might be possible if the government could include sex-disaggregated data in the NIRF, such as: numbers on women doctorates, postdoctorates, women faculty in science and engineering in an institute, data on how many men and women are there at various positions in the organizational hierarchy in an institute, and on female and male recipients of external funding. Further, we need an autonomous body that can collate data gathered from various sources, such as the Census, NIRF, and DST for the purpose of understanding various dimensions of S&T education and personnel.

Work-life balance

A lack of work-life balance structures is creating a debilitating impact on women scientists (and women professionals in general) as they consolidate the assumptions of a dual burden on women. The structures are required not only for a better work-life balance for women and men, but also for creating a socio-psychological climate in which it might be assumed by the recruiters and the senior colleagues that women can take care of family (since the assumption that women are central to domestic roles is not likely to change in the near future).

As Ramadoss and Rajadhyaksha (2012) remark, by itself, multiple roles are not stressful unless one finds it difficult to manage those roles. Studies find that families are rejecting the traditional gendered dichotomy of public-private sphere, and that men and women in dual-career families are committed to both work and family roles, although working women demonstrate greater egalitarian gender role ideology than men (Rajadhyaksha and Velgach 2015). However, there is a structural lag and policies and workplaces are changing at a much slower pace than the familial space. In fact, men receive greater support from colleagues, bosses, and extended family members than women because of the structural lag and the traditional mindset that women do not need support (Ramadoss and Rajadhyaksha 2012). Increasingly, a crèche is required not only for women and men professionals but also for doctoral, postdoctoral researchers, and for those attending conferences. Other support structures could include age-based support for childcare and commuting facilities for scientists residing outside campuses.

It is imperative that various provisions such as childcare leave are made available to men and that the structures (crèche) and family-related leave are used by men also. Not only the organizational support for family roles is lacking but also, as indicated in the earlier chapter, the available structures and policies are perceived to be facilities for women, and flexi-leave or the paternity leave is seldom availed by men justifying the masculine culture of long hours. Studies show that when organizational culture aggressively encourages employees to use family-friendly policies by tying executives' compensation to the usage and results of work/life programs, employees take advantage of them (Ridgeway and Correll 2000). In Sweden (ranked first among the Organization for Economic Cooperation and Development or OECD countries in labor force participation rate), laws give parental leave of 480 days per child, of which 384 days before the child turns four years old and the rest before the child is 12 years old or completes the fifth grade. Both parents can share these days, *although 90 are reserved for each of the parent.*[13] Indian childcare leave is much more generous, but is meant for women and single men only.

Dealing with biological differences

India has one of the most generous paid maternity leave rules in the world. In the US, the lack of paid maternity leave is believed to be one of the reasons for a lack of women in science (Abad-Santos 2013). Pregnancy is an obvious biological difference and logically, it should not entail a loss in terms of professional development.

Some enlightened suggestions and measures have been put forth lately to avoid professional loss resulting in this period. For instance, Ananth (2014) suggests that for women scientists taking maternity leave, a 3/2 multiplicative factor to the work she completed during the period should be applied when considering evaluation. IISC, since 2017, provides a time extension in promotion and evaluation to the women faculty after maternity leave.

As shown earlier, the maternity leave in India is viewed as a loss to the workplace by colleagues and the bosses. This is also evidenced in the government efforts to compensate companies for the loss of part of the period. However, in academia/research labs, it is difficult to compensate and counter the feeling (of colleagues) that women 'take-off' after every few years. This requires social change, such as change in gendered expectations, which in turn requires normative change in socialization patterns and acceptance of women's equal right to public spaces.

Integrating caste and gender

As the discussions on caste and gender in this work show, the Dalit/Adivasi women face both caste and gender discrimination, and the stereotypes and biases against the lowest castes conflate the gender biases resulting in a severely harmful impact on their life chances.

There is a near absence of studies and public discourse on the caste question in science (Thomas 2020). Dalit/Adivasi women in STEM academia are underresearched and undersurveyed, which prevents recognition of discrimination against the group. There are complaint cells in the institutes, SC/ST cells in some institutes and a commission at the national level dealing with caste-based harassments. Yet, given the magnitude of the problem, the fight against caste and gender requires innovative methods and substantial will. The suicide of a Dalit student at IIT Bombay in February 2023 prompted the institute to issue guidelines (with severe consequences for those violating them) to prohibit caste discrimination that include prohibiting students from asking each other anything that might reveal caste, such as JEE (Advanced) rank, GATE score, to crack casteist jokes, share casteist messages, etc.

Caste discrimination in academia could be structural, inadvertently constructed in policies and practices through caste-based categorizations (Rao 2013). There is a need to apply a caste and a gender lens to all policies and practices. Cells within the institutes that monitor SC/ST discrimination should be strengthened and collaborate with anti-sexual harassment cells to deal with the complaints of Dalit/Adivasi female students. At the institute level, programs should be devised that monitor and support the careers of the female Dalit/Adivasi students, and these might be helpful in creating programs at a larger scale. Other possible measures could be: devising creative ways to educate the upper castes on how caste could create barriers in STEM careers, tapping into the pool of sensitive upper caste students and faculty to raise awareness about the existence of caste barriers within the institute and to find solutions for creating an egalitarian research environment by consciously inviting successful Dalit/Adivasi scientists for talks/panel discussions.

Best practices elsewhere

Throughout the world, several programs have been initiated to highlight the issue of gender equality in science. This has resulted in improvements, although there is still a great deal to be achieved (Shen 2013). There is much to learn from these programs and adapt them to our national context. Some examples are given below.

Gender mainstreaming

Gender mainstreaming was endorsed at the World Conference on Women in Beijing in 1995 (UN 2002). It emerged worldwide because of the dissatisfaction with the strategies that focused on women, and because there was 'recognition that inequality between women and men was a relational issue and that inequalities were not going to be resolved through a focus only on women ...' (UN 2002: 9).

This approach suggests that all legislation, policies, and implementation should be assessed in terms of implications for women and men, that women's and men's concerns and experiences should be integral for any design, implementation, monitoring, and evaluation of policies and programs in all spheres 'so that women and men benefit equally. ... The ultimate goal is to achieve gender equality' (UN 2002: 1).

Gender mainstreaming could be implemented at all levels, including government policies as well as in an organization's programs (e.g., EIGE's Gender Mainstreaming Guide for Organisational Change).[14] It is generally based on the idea that there is no gender-neutral policy, and that the implemented actions have a positive or negative impact on people's lives, that is, women, men, and the LGTBQIA+ community (Parejo and Radulovic 2023). The concept is accepted as a significant component to the policymaking process internationally. It also refers to the inclusion of gender expertise into the policy process (Parejo and Radulović 2023).

Gender mainstreaming (or GM) in the context of institutional transformation aims at the structures of the organization, rather than targeting the individual women as 'lacking' (and thus responsible for their own discriminated position). In practice, it requires all staff to have knowledge of how to consider gender equality in their everyday work practices (Eveline et al. 2010: 241–242). It includes, for example, training for recruitment committees as a way to remove barriers to women gaining professorships (Winchester et al. 2006). GM may result in a new governance structure that integrates all relevant strategic and operational actors and units, the establishment of transparent, structured, and formalized procedures, etc.[15]

GM has been carried out in a variety of organizations from multilaterals, to governments, to small and large non-governmental organizations. Nevertheless, there are reviews suggesting many failures of GM programs. The failure of the strategy owes mainly to losing sight of transformative agendas in trying to make the approach palatable to the organizations, technical issues such as lack of integration of gender perspective and analysis in plans and monitoring, deep-seated biases, complex power structures, and overabsorption in organizational dynamics of GM losing sight of the results achieved on the ground. However, it has brought indirect benefits through increasing the theorizing and understanding of the gendered nature of organizational practices, procedures, routines, and cultures (Milward et al. 2015).

GATI in India aims to enhance gender equality through mainstreaming gender inclusion policies across all institutes (see the GATI website). However, a successful implementation of gender mainstreaming requires the commitment of senior management and the establishment of effective accountability mechanisms (UN 2002).

Targeting academic institutes

In 2001, the National Science Foundation (NSF) in the US launched the ADVANCE program to catalyze change that aimed to transform academic environments in ways that enhanced the participation and advancement of women in science and engineering. Prior to the NSF ADVANCE program, most efforts at increasing the numbers of women faculty in academic STEM focused on 'fixing the women' (a belief that women lacked in skills or qualifications compared to men). The ADVANCE program focused on 'fixing the institutions' (a belief that the inequality is systemic) instead (Rosser 2004). The NSF Awards support comprehensive programs for institution-wide change. A seminal program within the ADVANCE portfolio is the Institutional Transformation Grant, which provides large scale, multiyear funding to institutions of higher education (IHE) focused on fundamental organizational changes that should broaden participation of women and foster gender equity in STEM academic careers.[16]

Since 2001, NSF has invested $270 million to support ADVANCE projects at more than 100 institutions of higher education (IHE) and STEM-related not-for-profit organizations.[17] Some of the strategies used by the IHEs include increasing awareness of, and reducing, implicit gender bias among members of faculty search committees through workshops (after data collection at the institution), department-level climate change using the dual-agenda approach, attempts to boost careers from associate to full professor through the development of a demographic database to track the gender distribution among faculty ranks over time, attempts to raise awareness and action through workshops, panel discussions and departmental meetings, and policy changes such as clarifying post-tenure review, addressing issues of isolation by stimulating research collaborations through seed money, and travel grants for woman-led research project teams, interdisciplinary knowledge exchanges, cross-sector research showcases, and an informal Research Café (Furst-Holloway and Miner 2019).

There are evaluation/accreditation programs, such as Athena SWAN (Scientific Women's Academic Network) Charter in Great Britain and SAGE (Science Australia Gender Equity) in Australia, where the institutes that sign the charter commit themselves to promoting and supporting gender equality for women. Such institutes apply for awards which have been linked to funding. Drawing inspiration from Athena SWAN, the GATI program in India, as discussed earlier, was initiated in 2020.

Impact evaluation

Policy evaluation is essential to analyze whether policies are able to fulfill their objectives and to take corrective measures to further the goal of gender equality. There are two aspects to it, the first one is examining the impact of all policies from a gender perspective. Gender mainstreaming encompasses the evaluation of policies from a gender equality standpoint. Several EU Member States are forerunners in gender-based public policy analysis. For example, in the Netherlands, a framework for gender (emancipation) impact assessment was introduced in 1992 (Parejo and Radulović 2023).

The second aspect is the evaluation of policies aimed at gender equality. Countries and international organizations undertake reviews of their gender equality policies e.g., Irish Aid (2010) or evaluation by the World Food Programme (2020). Most evaluation plans and procedures target implementation of gender equality schemes of the institutes or organizations. For example, at the initiative of the Australian government, a guide was brought out to provide advice on how to evaluate STEM gender equity programs (Kingsley 2020). However, assessing the impact of gender equality is acknowledged as extremely complex due to the complexity of programs, a large number of factors that interact in generating impacts, and the involvement of a large number of variables, as for instance in Athena SWAN (Schmidt and Ovseiko 2020). UNESCO provides a STEM and Gender Advancement (SAGA) toolkit to aid nations in their evaluation of policies affecting gender equality in STEM and to enable evidence-based policymaking (UNESCO 2017).

Given the large range of programs by the Indian government for women in STEM, some of them unique in the world, it seems imperative to devise a framework of our own to evaluate those and to have a dedicated team for this purpose. Without a scientific evaluation system, the gains of the programs would not be commensurate with the time, resources, and effort expended on them.

Social change as enabler of gender equality

Challenging gender inequality requires change in gendered expectations, norms, and attitudes. Gender attitudes, even those rooted in centuries-old cultural norms, are amenable to change. For example, reserving seats for women in village councils has helped curtail negative stereotypes about women as leaders in India (Beaman et al. 2009). There are evidences of programs designed for school children having positive outcomes on attitudes towards girls and women (e.g., Dhar, Jain, and Jayachandran2022).

Indian society exhibits continuity and change. As shown in the earlier chapters, the defining femininity of teens carries attributes of women as family-oriented and sexually chaste, typical of middle-class families (Diwakar 2016). The son preference and differential socialization of boys and girls create male privilege and masculinity in society and institutions of education. Construction of self-assured masculinity is impeded due to inhibited interaction (Iyer 2014). These attributes translate into male entitlement and women's marginalization in public spaces.

However, the Indian social system is also undergoing dynamic change, and with it the position of women is getting transformed. This is evident in the substantial rise in the women's proportion in STEM education since the 1990s, and an engineering education for women being regarded as prestigious, with the proportion of female enrollments in UG in engineering rising from 16% in 1995 (Banerjee and Muley 2008) to 29% in 2021–2022 (AISHE 2021–2022). The implementation of neo-liberal policies from the 1990s has brought considerable changes in the cultural practices of the Indian middle classes, which appear to have affected women and gender relations (Fernandes 2006). For example, media portrayal of women indicates fewer restrictions on women's movements and dress (Dewey 2008), there is a greater acceptability of women in the public sphere, families headed by young couples display greater egalitarian attitudes and are on the rise, dual-earning couples are becoming increasingly common in urban India (Ramadoss and Rajadhyaksha 2012), and living apart with no one having to relocate is gaining prominence among the educated urban upper-class couples (Gangopadhyay 2023). This transitional milieu has raised the possibility of multiple masculinities and femininities, thus challenging the prevalent notions of masculinity at the research institutes. Younger men or those with working wives are more sympathetic to the dual burden concerns of women (Gupta 2016). While a woman's role in the family continues to be regarded as pivotal by the society, including by women scientists themselves, change is witnessed in terms of greater family support and a higher determination among women to pursue a career (Gupta 2016).

There are evidences of a growing interest regarding gender equality issues among the male scientists, as discussed next.

Men as champions of change

The struggle against gender inequality inevitably requires the involvement of men. The Beijing Platform for Action in 1995 declared that women's concerns could only be addressed 'in partnership with men'.[18] However, men in

STEM, as noted above, are usually excluded from the discourse on gender equality since gender issues are perceived as matters related to women.

Discussions of gender often create discomfort among men because gender is generally perceived as discrimination against women and simultaneously implies privileges of men. Studies on the psychology of prejudice point out that discrimination and privilege are two sides of the same coin (Kite and Whitley 2016). Those with privilege often become annoyed when confronted by the fact of their privilege, as it seems to negate the hard work and their own efforts in their accomplishments (Ferber 2012). However, as Johnson (2006: 21) remarks, the existence of privilege does not mean that the person did not do a good job or does not deserve credit for it; rather, it means that he/she is getting something that others are not, even though they are similar in every respect except for the social category to which they belong. Hence, access to privilege does not determine outcomes, but it is an asset that makes positive outcomes more likely. Thus, stereotypes in society and masculinity in science privilege men and make their success more likely than that of women, even though the latter may be similar in talent, ability, and aspirations. Hence, the issue is not of men versus women but of an unequal social, institutional, and organizational system that creates disadvantages for women professionals in science and technology.

Men were the pioneers in championing the cause of women in Indian history. The Indian social reformers in the colonial period who fought against the prevalent social evils against women such as Sati, *purdah*, and championed widow remarriage and women's education, were male. Gandhi played a major role in encouraging women to come out of their homes in protest against the British, and thereby created visibility for women in public spaces. The views of the reformers and nationalists might seem limited when judged by today's aspirations and values; nevertheless, they were radical for those times. While women had to fight hard for suffrage rights in Britain and the US, the constitution-makers of India incorporated women's equal right to vote without any suffrage rights movements. However, such rights made little dent on the patrifocal ideology, particularly on the norm of the 'public-private' dichotomy.

In the post-independence period, the dominant reaction of men to feminism in India has been one of opposition (Choudhary and Baset 2018). For instance, in academia it has implied trivializing feminist concerns, theories, and methodologies (Tellis 2000). However, since the 1990s, pro-feminist men's organizations have emerged to address violence against women; the proliferation of NGOs has provided the 'conceptual infrastructure' necessary for such men's organizations (Choudhary and Baset 2018: 16). Two generations of male feminist scholars have been produced for whom feminism has been a major part of their formal intellectual training.

Since the start of the millennium, there has been a growing thinking on how men could also be a part of the solution, to engage with the policymakers to incorporate perspectives on masculinities in various gender-related initiatives in the country (Chatterji 2011). Men and masculinities in India form the subject of a UNESCO report (2021). It explores the cultural and historical roots of masculinity and emphasizes the need to engage and empower men and boys in India in order to challenge negative gender stereotypes and behaviors. Highlighting the manner in which cultures of masculinity shape men and the suffering due to power differentials among men (due to differences in caste, class, caste, region) can potentially provide inroads into modes of thinking that oppose patriarchal power (Srivastava 2018).

This work shows that there is not only dominance of men in numbers but also in positions of power in science institutions. Further, the workplace culture is 'masculine', that is, it exhibits practices and norms that are typically attributed to men. The predominance of certain masculine practices or the 'hegemonic masculinity' constructs women as subordinates (Connell 1995). However, studies of masculinity point out 'multiple masculinities' (Connell and Messerschmidt 2005), which provide multiple meanings of gender, thus challenging 'hegemonic masculinity'. Masculinity might not mean the same thing to all men (Coles 2009). It might, therefore, be hypothesized that by generating an alternative discourse of masculinity in STEM research, the current formulation of 'hegemonic masculinity' might be discredited, creating a greater inclusive culture.

There are already signs of an alternative discourse. For instance, there are men desirous of a better work-life balance (Gupta 2017) that might challenge organizational practices that construct dominant masculinity such as the long-hours work culture. Men in other professions also are increasingly seeking work-life balance from their workplace, wanting to participate in raising family and taking care of ailing parents, particularly among the younger generation (Singh 2018). In pursuing greater time for families and involvement with domestic duties, these men are challenging the gendered norm of separation of spheres.

Since the beginning of the millennium there has been a growing interest among the men scientists about the position of women. Malhotra (2018) notes more men participating in the discussion as shown by a significant increase in the proportion of articles authored by men in *Current Science*. Of the 23 articles between 2010 and October 2016, eight articles were with men as sole author/first author. An article, for instance, recognizes that 'the academic community has a skewed gender ratio, which gets worse when the focus shifts to leadership' (Ananth 2014: 1366); that one of the main reasons is the maternity leave which delays promotions and makes it harder to reach leadership positions. Among his suggestions include allowing a portion of research funds to be used for childcare, provision of crèche, and qualified childcare workers and a 30% quota for women on governing boards of all institutions.

Studies on men scientists are rare in India. But from the available evidence, there are some men who differ from the hegemonic viewpoint on the system's impact on women scientists (Gupta 2016). While most male scientists tend to deny that the organizational processes and practices create barriers for women, a few agree that the barriers, such as lack of acceptance of women in leadership positions and lack of godfathers are systemic (Gupta 2020). For instance, a male scientist giving his opinion on gender equality in his institute said:

> Gender equality means equality of opportunity. Such equality is not absent overtly. But I have seen very strange things here. People have stereotypes because of which they would like to keep women away from administrative responsibilities.
>
> *(Gupta 2020: 700).*

This statement implies an understanding that stereotypes operate at an organizational level that excludes women unfairly. Such men represent a version of masculinity, 'more open to equality with women' (Connell and Messerschmidt 2005: 853).

There are evidences of spouses of women scientists relocating to support their wives' careers (Gupta 2016), and of women scientists finding considerable support from their husbands that are scientists (Kurup et al. 2010). However, even though there might be a greater support at home in dual-career families (Ramadoss and Rajadhyaksha 2012), the workplace continues to create barriers, suggesting failure of the workplace to catch up with the social changes. For example, Aruna Dhathathreyan, Raman Research fellowship awardee and Professor Emeritus Scientist at a CSIR lab (Central Leather Research Institute) writes in Lilavati's Daughters (Godbole and Ramaswamy 2008: 104):

> To be accepted as a scientist who happens to be a woman is still an uphill task. ... A woman is expected to be docile and not ask too many questions.

However, Dhathathreyan attributes her survival in science to her husband, also a chemist, sharing in child-rearing.

Through their participation in challenging gender inequality, there emerge the possibilities of men 'undoing gender' (Kelan 2010) at the workplace. For instance, mentorship of female scholars by male advisors and faculty members can help to promote women's careers in STEM (Blake-Beard et al. 2011). Men insisting on female members in scientific panel discussions or workshops, supporting female faculty members or female scholars in cases of harassment are examples of such challenges by men. Such support of male scientists is visible through social media, as for instance, on X (formerly Twitter), some men scientists reacting to the all-male panel of scientists or to a lack of women in departments of elite institutes.[19]

To enhance the awareness among men about gender issues and to create greater support for gender equality, more research studies and data need to be generated and policies need to be gender sensitive and not merely women-centric. Further, introducing an optional course on 'gender, science, and technology' in engineering/science institutes might be considered. Cultivating gender awareness (through such courses) in the institutes that train future generations of scientists/engineers might have a ripple effect.

New enablers: social media and private sector

There are social media platforms, private industries, and the NGOs supporting and encouraging women in STEM. The social media platforms aid in voicing the concerns and issues of such women. The Life of Science (TLoS), started by women science communicators in 2016, collected valuable narratives on the struggles of women scientists across the country, continuing their work in this direction; BiasWatchIndia, an online platform initiated by women scientists, documents gender representation in conferences, meetings and talks in India, and compares it to the current aggregate percentage (base rate) of women faculty in Indian scientific institutions.

To encourage women in STEM education and jobs, various initiatives are being undertaken by large scale international industries (Microsoft, Google, Adobe, L&T, Dr. Reddy's, etc.), such as hiring women in science who took career breaks, programs to help women to overcome professional challenges, to aid young girls to discover opportunities in ICT industry, or to provide scholarships to girls from weaker sections to pursue STEM education (UNESCO 2022). The Confederation of Indian Industry has formed the Indian Women Network that provides a structured network through which career women can meet to discuss and overcome challenges they might be facing at the workplace.

There are several initiatives of the civil society for underprivileged girls, such as Indian Girls Code, an initiative by Robotix, offering a free hands-on coding and robotics education program, which is recognized by Niti Aayog and UNESCO. Also, FAT (Feminist Approach to Technology), a not-for-profit organization, aims to reduce the gender gap in technology through teaching technical skills, encouraging innovation, conducting workshops/counseling, etc.

Pockets of promise for women with STEM education?

There are certain areas in STEM in India, which are gendered but compared to academia, they either exhibit more favorable circumstances for women or hold considerable promise. High-tech entrepreneurship, space science, and the aviation sector are examples of such areas. However, there is scant research in these fields.

Women-led tech start-ups in STEM academia

Academic incubators help to enhance female entrepreneurship, as noted in Chapter 3. These incubators leverage deep technology and through their pool of mentors, alumni networks, infrastructure, and funding (from government), work to de-risk entrepreneurship.

Given the patriarchal structure, female-founded startups in deep technology help to counter gender stereotypes. Experiences with academic incubators suggest positive results both for gender dynamics and enhancing an emergent entrepreneurial culture (Gupta and Etzkowitz 2021). Entrepreneurship is a source of empowerment for women by providing economic security, development of entrepreneurial behavior, and increased contributions to the family (Datta and Gailey 2012).

Entrepreneurship in the academic setting has emerged as a career choice in STEM and appears to be woman-friendly. A greater visibility for women founders in technology incubators at the academic institutes is likely to encourage women in this field. More research in this area will help to understand the problems and the measures required.

Women space scientists

Unlike academia, Indian space science has several women scientists in leadership roles, as noted earlier. This makes the space sector unique within STEM research. What makes this sector different? Vaid (2019) notes that due to the ISRO's emphasis on team effort, successes are never attributed to a single person, male or female. Thus, implying that women in the team are likely to be as much felicitated as men for the mission's success. In general, mixed-gender science teams have been found to be high performing groups (Yang et al. 2022). Further, institutes that fall under the ISRO and space funding agency (DOS) do not require application for grants, and as mentioned in Chapter 4, there is a significant correlation between women's high proportion as leaders and absence of requirement for application for grants (Godbole and Ramaswamy 2015). This indicates that there are gender issues in grants, and an absence of the 'grant' hurdle enables women's rise. This, in turn, suggests that greater transparency and efficiency in the system of grants are required.

In addition, scientists at the space organization surveyed (which has launched several successful missions), are viewed as 'scientist-workers' by the elite institutes such as the IISc (Arora 2022). Perhaps a lower status of the space organization in the hierarchy of research institutes works in favor of women as it does not attract the gender biases in the same way as the elite academic sciences. More research is required into what makes women scientists in space organizations successful: the role of the leadership, teams, structure, and culture of the organizations.

Female pilots

The aviation sector is another field that is witnessing an encouraging participation of women, offering a career route in STEM different from engineering/doctor/researcher. It cannot be ignored, as India has the highest proportion of female pilots in the world, which is 14.3% according to Civil Aviation Statistics (2023–2024),[20] almost three times the global average.[21] In the US, the share of female pilots was about 5% of the total in 2022.[22] To acquire a commercial pilot license in India requires grounding in STEM subjects, apart from the pilot training.

What has led to this high proportion, although sexism and misogyny at the workplace exist in this sector as in any other sector or country? Women were seldom considered for late night or early morning flights, until the last few years (Ramesh 2022). The growing aviation sector and the job opportunities in this sector, training facilities available to women, and equal pay are believed to be some of the factors (Shah 2018). Women are drawn to flying through an air wing of the National Cadet Corps, where students are trained to operate microlight aircraft (Saxena 2022). To make the expensive commercial pilot training more accessible to women, some state governments are subsidizing it, and some offer full scholarships for the course and help find placements. Private airlines, such as Indigo and Vistara, have work flexibility options, pregnancy leave, and creche in place; efforts to raise awareness and outreach in schools; and family support are also important factors (Saxena 2022).

The factors similar to those encouraging women in engineering, including high paying career options, the urban, middle-class families increasingly accepting of career choices of their daughters, and women's own determination to pursue careers of their choice could be encouraging women to take up careers in aviation. More studies are needed to analyze women's participation in this sector, which might hold lessons for other areas in STEM.

Critical areas for research on gender in scientific research

To create innovative solutions to the problem of underutilization of women's talents in science, one of the primary requirements is more research on women in STEM. For this, it will be helpful to have research centers in the IITs and premier institutes of S&T for an exclusive study of women in science. The findings of such centers/scholars should be in the public domain and accessible to the policymakers. Specifically, research studies are required urgently in the following areas: one, the study of recruitment and promotion procedures, the importance of research productivity, networking, and visibility in construction of 'merit'; two, the study of male and female students in master's, doctoral, and postdoctoral STEM fields regarding their career plans, higher studies abroad, and bottlenecks. Longitudinal studies from Master's level in this regard will help to trace the nature of career paths of men and women and the role of gender in careers; three, the study of intersectionality of caste and gender to understand caste disadvantage in academia; four, a social network analysis to understand the nature of exclusion from networks, facilitators, or bottlenecks in collaboration for women; five, comprehensive statistical studies on research productivity and its relation with marriage, parenthood, and age will help in understanding if men and women have different periods at which they produce maximum/minimum research; and six, historical and descriptive studies regarding how India's post-colonial structures of science and its third-world status impact women scientists in terms of opportunities for research.

These and various other aspects pointed out in the book need to be further researched to understand the nature of the issues and to find solutions to gender inequality in STEM research.

The final argument

This book analyzed how, at various stages and in various fields, women in STEM continue to be disadvantaged due to social and institutional structures. Although the women are overwhelmingly from upper-caste, middle-class backgrounds, the latter fails to provide the cultural and social capital that is available to the men of their social background for success, owing primarily to gender. The talent of the lower caste women continues to remain hidden. In its inability to use the potential of its women in STEM fully, India fails to empower itself socially and economically.

There are several indications that attitudes regarding the subject of 'gender and STEM' have not been static. Within a span of the last two decades or so, we have traversed the distance from being oblivious to gender inequities to an emergent awareness, leading to denial among some, but also an increasing acceptance among the scientific community; from government's focus on numbers to gender equity and gender equality; and from near silence to an active engagement of women scientists and emergence of other enablers on gender issues.

However, there is still a need for greater awareness of how gender inequality operates in STEM. We have many enlightened policies, but much more needs to be done in terms of implementation. Finally, there is a need to recognize that the relationship between gender and STEM is not about women, but about the structures and the socio-cultural environment that constrain women from reaching their full potential.

Notes

1 Shubha Tole, HT, April 1, 2023: https://www.hindustantimes.com/opinion/addressing-the-gender-gap-in-stem-fields-why-changing-exam-formats-is-not-enough-101680350856820.html.
2 Initiatives: https://www.ias.ac.in/Initiatives/Women_in_Science/Initiatives.
3 ‘Women in Science’ Please see: https://www.ias.ac.in/Initiatives/Women_in_Science/; Inter-Academy Panel for Women in STEMM (IAP-WiSTEMM): https://wistemm.in/.
4 Standing Committee: http://www.dst.gov.in/sites/default/files/Standing-Committee-OM-Composition.pdf
5 As on the DST website: https://dst.gov.in/sites/default/files/ADVERTISEMENT-Mobility.pdf.
6 Recommended in the Women in STEM workshop at New Delhi in 2015, Page 9: https://www.britishcouncil.in/sites/default/files/report_-_women_in_stem.pdf; also see: https://www.ndtv.com/science/women-scientists-seek-parity-some-want-equal-share-for-men-in-childcare-leave-5198386; by court: https://www.livelaw.in/high-court/calcutta-high-court/child-care-leave-should-be-given-to-both-male-and-female-govt-employees-family-responsibilities-must-be-shared-by-mother-father-calcutta-high-court-266679
7 National Guidelines for Crèches Set Up Under the Maternity Benefit Act: https://www.livelaw.in/law-firms/articles/national-guidelines-crches-maternity-benefit-act-142685.
8 Please see https://www.downtoearth.org.in/blog/science-technology/did-the108th-indian-science-congress-uphold-its-theme-of-women-empowerment--87083
9 GATI website: https://www.gatiresources.in/gati-framework/gati-charter
10 For a definition of gender equity see https://eige.europa.eu/thesaurus/terms/1175
11 For a definition of gender parity see: https://eige.europa.eu/publications-resources/thesaurus/terms/1258
12 For a definition of gender equality see: https://eige.europa.eu/publications-resources/thesaurus/terms/1059
13 Sweden — Parental benefits and benefits related to childbirth. European Commission. Please see: https://ec.europa.eu/social/main.jsp?catId=1130&intPageId=4808&langId=en
14 Gender Mainstreaming: A Guide to Organisational Change: https://eige.europa.eu/gender-mainstreaming/toolkits/gender-institutional-transformation/gender-mainstreaming-guide-organisational-change
15 EIGE. Gender Mainstreaming. Available at: https://eige.europa.eu/gender-mainstreaming
16 Please see the NSF website: https://www.nsf.gov/funding/pgm_summ.jsp?pims_id=5383
17 ADVANCE at a Glance: https://www.nsf.gov/crssprgm/advance/.
18 Beijing Declaration and Platform for Action, 1995: https://www.un.org/womenwatch/daw/beijing/pdf/BDPfA%20E.pdf

19 For e.g., Anant Bhan, @AnantBhan, Sept 7, 2019. Available at: https://twitter.com/AnantBhan/status/1170358094011846657
20 Handbook on Civil Aviation Statistics (2023–24): Chart 22. https://www.dgca.gov.in/digigov-portal/?page=4252/4205/sericename
21 Women pilot statistics: https://www.icao.int/Newsroom/Pages/ICAO-releases-survey-data-on-status-of-global-aviation-gender-equality.aspx; https://pilotinstitute.com/women-aviation-statistics/; Also see: https://www.statista.com/chart/27953/percent-share-of-female-commercial-airline-pilots-by-country/
22 'Women airline pilots': https://centreforaviation.com/analysis/reports/women-airline-pilots-numbers-are-growing-but-still-a-pitiful-percentage-655755.

References

Abad-Santos, Alexander. (2013). The First Step to Having More Women in Science Is Maternity Leave. *The Atlantic*, October 7, 2013. Available at: https://www.theatlantic.com/politics/archive/2013/10/first-step-having-more-women-science-maternity-leave/310246/. Accessed on September 6, 2024.

AISHE 2021–2022 *All India Survey on Higher Education 2021-2022*. Government Of India. Ministry of Education, Department of Higher Education: New Delhi. Available at: https://aishe.gov.in/aishe-final-report/. Accessed on September 6, 2024.

Ananth, Sudarshan. (2014). Women leaders in Indian science. *Current Science* 107(9): 1366.

Arora, C. (2022). Sustaining Science, Becoming Respectable. *Seminar: The Monthly Symposium* 760(December 2022: Women in Science): 26–32. Available at: https://www.india-seminar.com/semframe.html. Accessed on September 6, 2024.

Ashok, S. 2018. *Few Women in Science, Fewer on Stage of Science Congress to Speak about it*. March 19, 2018. Available at: https://indianexpress.com/article/india/few-women-in-science-fewer-on-stage-of-science-congress-to-speak-about-it-5102624/. Accessed on September 6, 2024.

Banerjee, R., and Muley, V.P. (2008). *Engineering Education in India*. Draft final report. Sponsored by Observer Research Foundation; Energy Systems Engineering, IIT Bombay Powai: Mumbai. Available at: https://www.academia.edu/28572868/ENGINEERING_EDUCATION_IN_INDIA. Accessed on September 6, 2024.

Batchelor, R. L., Ali, H., Gardner-Vandy, K.G., Gold, A.U., MacKinnon, J.A., and Asher, P.M. (2021). Reimagining STEM workforce development as A braided river. *Eos* 102(19th April 2021). https://doi.org/10.1029/2021EO157277. Accessed on September 6, 2024.

Beaman, L., Chattopadhyay, R., Duflo, E., Pande, R., and Topalova, P. (2009). Powerful Women: Does Exposure Reduce Bias? *The Quarterly Journal of Economics* 124(4): 1497–1540.

Bhopal, K., and Henderson, H. (2019). Competing Inequalities: Gender Versus Race in Higher Education Institutions in the UK. *Educational Review* 73(2): 153–169.

Blake-Beard, S., Bayne, M.L., Crosby, F.J., and Muller, C.B. (2011). Matching by Race and Gender in Mentoring Relationships: Keeping Our Eyes on the Prize. *Journal of Social Issues* 67(3): 622–643.

Cabrera, S.F., and Thomas-Hunt, M.C. (2007). "Street Cred" and the Executive Woman: The Effects of Gender Differences in Social Networks on Career Advancement. In S.J. Correll (ed), *Advances in Group Processes: Vol. 24. Social Psychology of Gender*, Emerald: New York, 123–147. Available at: https://ecommons.cornell.edu/items/eceb0818-fc03-4ba5-be45-fb1b0e4fda38. Accessed on September 6, 2024.

Ceci, S.J., and Williams, W.M. (2015). Women Have Substantial Advantage in STEM Faculty Hiring, Except When Competing Against More-Accomplished Men. *Frontiers in Psychology* 6: 1532. https://doi.org/10.3389/fpsyg.2015.01532
Chander, M. (2021). Deepa Mohanan Won a Rare Battle Against Academic Casteism, But It Wasn't Because of the Law. *Article 14*, November 19, 2021. Available at: https://article-14.com/post/deepa-mohanan-won-a-rare-battle-against-academic-casteism-but-it-wasn-t-because-of-the-law-619716f6ed5d2. Accessed on September 6, 2024.
Charlesworth, Sara, and Baird, M. (2007). Getting Gender on The Agenda: The Tale of Two Organisations. *Women in Management Review* 22(5): 391–404.
Chatterji, S. (2011). Back to Family. *India Together.* December 30, 2011. Available at: http://www.indiatogether.org/praajak-children. Accessed on September 6, 2024.
Checchi, Daniele, and Cicognani, Simona, and Kulic, Nevena. (2015). *Gender Quotas or Girls' Networks? Towards an Understanding of Recruitment in the Research Profession in Italy.* Quaderni – Working Paper DSE N° 1047. Available at: http://dx.doi.org/10.2139/ssrn.2707577. Accessed on September 6, 2024.
Choudhary, R., and Baset, Z.A. (2018). Introduction. In R. Choudhary and Z.A. Baset (eds), *Men and Feminism in India*, 1st edition (Hardback), Routledge: India, 1–18.
Chowdhury, S.R. (2017). For the First Time, Three Science Journals are Publishing Special Issues Centred on Women Scientists. *Scroll.in*, March 21, 2017. Available at: https://scroll.in/article/831460/for-the-first-time-three-science-journals-are-publishing-special-issues-centred-on-women-scientists. Accessed on September 6, 2024.
Coles, T. (2009). Negotiating the Field of Masculinity: The Production and Reproduction of Multiple Dominant Masculinities. *Men and Masculinities* 12(1): 30–44.
Connell, R.W. (1995). *Masculinities.* Polity Press: Cambridge.
Connell, R.W., and Messerschmidt, James W. (2005). Hegemonic Masculinity: Rethinking the Concept. *Gender and Society* 19(6): 829–859.
Datta, P.B., and Gailey, R. (2012). Empowering Women through Social Entrepreneurship: Case Study of a Women's Cooperative in India. *Entrepreneurship Theory and Practice 36*(3): 569–587.
Deutsch, F.M. (2007). Undoing Gender. *Gender and Society* 21(1): 106–127.
Dewey, S. (2008). *Making Miss India Miss World: Constructing Gender, Power, and the Nation in Postliberalization India.* Syracuse University Press: Syracuse.
Dhar, D., Jain, T., and Jayachandran, S. (2022). Reshaping Adolescents' Gender Attitudes: Evidence from a School-Based Experiment in India. *American Economic Review* 112(3): 899–927. https://doi.org/10.1257/aer.20201112
Diwakar, Vaishali. (2016). 'It's Complicated': The Construction of Indian Middle-Class Teens in Social Media. *Journal of Creative Communications* 11(2): 161–182.
Dogra, A. (2016). Mayurika Starts a Daycare Revolution. *The Life of Science.com.* Published on September 12, 2016. Available at: https://thelifeofscience.com/2016/09/12/mayurika-starts-a-daycare-revolution/. Accessed on September 6, 2024.
Dogra, A., and Jayaraj, N. (2023). *Lab Hopping. A Journey to Find India's Women in Science.* Penguin: India.
DST (2010). *DST Task Force Report 2009–2010.* Report of the DST Task Force for Women in Science. Available at: https://www.ias.ac.in/public/Resources/Initiatives/Women_in_Science/taskforce_report.pdf. Accessed on September 6, 2024.

ECU (2014). *Evaluating the Effectiveness and Impact of the Athena SWAN Charter. Executive Summary.* Equality Challenge Unit. May 2014. Loughborough University. Available at: https://s3.eu-west-2.amazonaws.com/assets.creode.advancehe-document-manager/documents/ecu/evaluating-the-effectiveness-and-impact-of-the-athena-swan-charter_1579079953.pdf. Accessed on September 6, 2024.

EPW (2017). Review of Women's Studies. *Economic and Political Weekly* 52(17): 33–86.

Etzkowitz, H., Kemelgor, C., and Uzzi, B. (2000). *Athena Unbound: The Advancement of Women in Science and Technology.* Cambridge: Cambridge University Press.

Eveline, J., Bacchi, C., and Binns, J. (2010). Gender Mainstreaming Versus Diversity Mainstreaming: Methodology as Emancipatory Politics. In C. Bacchi and J. Eveline (eds), *Mainstreaming Politics: Gendering Practices and Feminist Theory*, Adelaide: University of Adelaide Press, 237–261.

Ferber, A.L. (2012). The Culture of Privilege: Color-Blindness, Postfeminism, and Christonormativity. *Journal of Social Issues* 68(1): 63–77.

Fernandes, L. (2006). *India's New Middle Class: Democratic Politics in an Era of Economic Reform.* University of Minnesota Press: Minnesota.

Furst-Holloway, Stacie, and Miner, K. (2019). Advancing Women Faculty in STEM: Empirical Findings and Practical Recommendations from National Science Foundation ADVANCE Institutions. *Equality, Diversity and Inclusion: An International Journal* 38(2): 122–130.

Gangopadhyay, R. (2017). *All These Happened to a Woman Scientist.* Project: DST WOS-A Scheme. Available at: https://www.researchgate.net/publication/317823465_All_these_happened_to_a_Women_Scientist._____Accessed on September 6, 2024.

Gangopadhyay, J. (2023). Living Apart Together (LAT): A New Family Form in Urban India. *Sociological Bulletin* 72(3): 332–348.

Gangopadhyay, R., and Das, B. (2016). DST WOS-A: The Scenario from Recipient's Perspective. *Current Science* 111(8): 1307–1309.

Gardner-Vandy, K.G., Batchelor, R.L., and Scalice, D.M. (2022). *Applying the Braided River STEM Workforce Development Model to Planetary Science and Astrobiology.* Advancing IDEA in Planetary Science, April 2022. Available at: https://ui.adsabs.harvard.edu/abs/2022LPICo2679.2035G/abstract. Accessed on September 6, 2024.

GATI Report (2022). *Gender Advancement for Transforming Institutions. A Baseline Study.* February 2022. British Council and DST. Available at: https://gati.dst.gov.in/resources/wspublic;jsessionid=CE0B091C8F66AAC5B2C520A68D567495/resource-material/GATI-Report-February_2023.pdf. Accessed on September 6, 2024.

Godbole, R. (2017). Women in STEMM: Involve the Institutions! *Current Science* 112(4): 671.

Godbole, R. M., and Ramaswamy, R. (2015). Women Scientists in India. In R.M. Godbole and R. Ramaswamy (eds), *Women in Science and Technology in Asia*, Indian Academy of Sciences: Bangalore, Annex 3, 67–84. Available at: https://www.ias.ac.in/public/Resources/Initiatives/Women_in_Science/AASSA_India.pdf._Accessed on September 6, 2024.

Godbole, R., and Ramaswamy, R., eds. (2008). *Lilavati's Daughters: The Women Scientists of India.* Indian Academy of Sciences: Bangalore.

Gupta, Namrata. (2007). How Critical Is Critical Mass? Implications of the Minority Status of Women in Science in India. *Current Science* 93(6): 766–768.

Gupta, Namrata. (2016). Perceptions of the Work Environment: The Issue of Gender in Indian Scientific Research Institutes. *Indian Journal of Gender Studies* 23(3): 437–466.
Gupta, Namrata. (2017). Gender Inequality in the Work Environment: A Study of Private Research Organizations in India. *Equality, Diversity and Inclusion: An International Journal* 36(3): 255–276.
Gupta, Namrata. (2020). Rationalizing Gender Inequality at Scientific Research Organizations, A Reproduction of The Indian Socio-Cultural Context. *Equality, Diversity and Inclusion, An International Journal* 39(6): 689–706.
Gupta, N., and Etzkowitz, H. (2021). Women Founders in A High-Tech Incubator: Negotiating Entrepreneurial Identity in The Indian Socio-Cultural Context. *International Journal of Gender and Entrepreneurship* 13(4): 353–372.
Heilman, M.E. (1996), Affirmative Action's Contradictory Consequences. Journal of Social Issues, 52: 105-109.https://doi.org/10.1111/j.1540-4560.1996.tb01852.x. Last accessed on November 16, 2024.
Heilman, M.E., and Alcott, V.B. (2001). What I Think You Think of Me: Women's Reactions to Being Viewed as Beneficiaries of Preferential Selection. *Journal of Applied Psychology* 86(4): 574–582.
Henningsen, L., Horvath, L.K., and Jonas, K. (2022). Affirmative Action Policies in Academic Job Advertisements: Do They Facilitate or Hinder Gender Discrimination in Hiring Processes for Professorships? *Sex Roles* 86: 34–48.
Husu, Liisa. (2000). Gender Discrimination in the Promised Land of Gender Equality. *Higher Education in Europe* 25(2): 221–228.
Ibanez, M., and Riener, G. (2018). Sorting through Affirmative Action: Three Field Experiments in Colombia. *Journal of Labor Economics* 36(2): 437–478.
INSA (2004). *Science Career for Indian Women: An Examination of Indian Women's Access to and Retention in Scientific Careers*. Indian National Science Academy (INSA): Delhi. Available at: https://www.ias.ac.in/public/Resources/Initiatives/Women_in_Science/report.pdf. Accessed on September 6, 2024.
Irish Aid (2010). *Annual Monitoring Report on Gender Equality – 2010*. Irish Aid, Government of Ireland. Available at: https://www.irishaid.ie/media/irishaid/allwebsitemedia/20newsandpublications/publicationpdfsenglish/annual-monitoring-report-on-irish-aid-gender-equality-policy-2010.pdf. Accessed on September 6, 2024.
Iyer, Padmini. (2014). *Negotiating Masculinities and Learning To 'Be A Man' at School in New Delhi, India*. A paper presented at the 2nd MenEngage Global Symposium, New Delhi, November 11, 2014. Available at: http://sro.sussex.ac.uk/55744/1/Iyer_%282014%29_Negotiating_masculinities_and_learning_to_%27be_a_man%27_at_school_in_Delhi.pdf. Accessed on September 6, 2024.
JAB (2017). *Highlights of the Recommendations of the JAB Sub-Committee for Increasing Female Enrolment in B.Tech. in IITs*. June 14, 2017. Available at: http://students.iitmandi.ac.in/~tag/doc/Report_Highlights_JAB_SC_Female_BTech_14Jun17.pdf. Accessed on September 6, 2024.
JEE Report (2024). JEE (Advanced) Report 2024. IIT Madras. Available at: https://jeeadv.ac.in/reports/2024.pdf. Last accessed on November 16, 2024.
Jayaraj, Nandita. (2021). *Sexual Harassment in Indian Science*. The Life of Science.com, December 31, 2021. Available at: https://thelifeofscience.com/2021/12/31/sexual-harassment-in-indian-science/. Accessed on September 6, 2024.
Johnson, A.G. (2006). *Privilege, Power, and Difference*. 2nd edition. McGrawHill: New York. Available at: https://blogs.lanecc.edu/engaging-diversity/wp-content/uploads/sites/76/2015/02/Privilege-Oppression-and-Difference.pdf. Accessed on September 6, 2024.

Jolly, P. (2022). Catalysing Change for Gender Advancement in Science. *Seminar: The Monthly Symposium* 760(December 2022: Women in Science): 33–38. Available at: https://www.india-seminar.com/semframe.html. Accessed on September 6, 2024.

Kelan, E.K. (2010). Gender Logic and (Un)Doing Gender at Work. *Gender, Work and Organization* 17(2): 174–194.

Kingsley, I. (2020). *Evaluating Stem Gender Equity Programs: A Guide to Effective Program Evaluation.* Women in STEM Ambassador. An Australian Government Initiative. Available at: https://www.unsw.edu.au/content/dam/pdfs/science/general/research-reports/2024-05-women-in-stem/2024-05-National-Evaluation-Guide-Oct-2022.pdf. Accessed on September 6, 2024.

Kite, Mary E., and Whitley, B.E. Jr. (2016). *Psychology of Prejudice and Discrimination.* 3rd edition. Routledge: New York.

Kurup, A., and Maithreyi, R. (2011). Beyond Family and Societal Attitudes to Retain Women in Science. *Current Science* 100(1): 43–48.

Kurup, Anitha, Maithreyi, R., Kantharaju, B., and Godbole, Rohini. (2010). *Trained Scientific Women Power: How Much Are We Losing and Why?* IAS–NIAS Research Report. Available at: http://eprints.nias.res.in/142/1/IAS-NIAS-Report.pdf. Accessed on September 6, 2024.

Llorens, A., Tzovara, A., Bellier, L., Bhaya-Grossman, I., and Bidet-Caulet, A. (2021). Gender Bias In Academia: A Lifetime Problem That Needs Solutions. *Neuron* 109(13): 2047–2074.

Malhotra, C. (2018). The Malaise of Under-Representation of Women in Science: The Indian Story. *Current Science* 115(9): 1714–1723.

Mehta, A.K. (2020). Union Budget 2020–2021: A Critical Analysis from the Gender Perspective. *Engage-EPW* 55(16): 1–22. Available at: https://www.epw.in/engage/article/union-budget-2020-21-from-gender-perspective#:~:text=Gender%20Budget%202020%E2%80%9321%20Allocations,14%2C893.40%20crore%20in%20Part%20B. Accessed on September 6, 2024.

Milward, K., Mukhopadhyay, M., and Wong, Franz F. (2015). Gender Mainstreaming Critiques: Signposts or Dead Ends? *IDS Bulletin* 46(4): 75–81.

Moss-Racusin, C.A., Pietri, E.S., van der Toorn, J., and Ashburn-Nardo, L. (2021). Boosting the Sustainable Representation of Women in STEM With Evidence-Based Policy Initiatives. *Policy Insights from the Behavioral and Brain Sciences* 8(1): 50–58.

Nielsen, Mathias W. (2016). Limits to Meritocracy? Gender in Academic Recruitment and Promotion Processes. *Science and Public Policy* 43(3): 386–399.

Niti Ayog (2017). *Final Report Status of Women in Science among Select Institutions in India: Policy Implications. Appendix.* SSESS, Kolkata. Available at: https://www.niti.gov.in/sites/default/files/2019-01/Appendix_Women_In_Science_SSESS.pdf. Accessed on September 6, 2024.

Ovseiko, P.V., Chapple, A., Edmunds, L.D., and Ziebland, S. (2017). Advancing Gender Equality Through the Athena SWAN Charter for Women in Science: An Exploratory Study of Women's and Men's Perceptions. *Health Research Policy and Systems* 15(12). https://doi.org/10.1186/s12961-017-0177-9. Accessed on September 6, 2024.

Padavic, I., Ely, R., and Reid, E. (2019). Explaining the Persistence of Gender Inequality: The Work–Family Narrative as a Social Defense against the 24/7 Work Culture. *Administrative Science Quarterly*, February 1, 2014: 1–51. Available at: https://doi.org/10.1177/0001839219832310. Accessed on September 6, 2024.

Parejo, V.H., and Radulović, B. (2023). Public Policies on Gender Equality. In D. Vujadinović, M. Fröhlich, and T. Giegerich (eds), *Gender-Competent Legal Education.* Springer Textbooks in Law, Springer: Cham, 405–428. https://doi.org/10.1007/978-3-031-14360-1_12. Accessed on September 6, 2024.

Powell, Stina. (2018). Gender Equality in Academia. Intentions and Consequences. *International Journal of Diversity in Organisations, Communities and Nations. Annual Review* 1(18): 19–35.

Priyadarshini, S. (2023). Breaking Gender Stereotypes in Physics. *Nature India*, Podcast April 19, 2023. Available at: https://www.nature.com/articles/d44151-023-00040-0. Accessed on September 7, 2024.

Rajadhyaksha, U., and Velgach, S. (2015). What Is a Better Predictor of Work-Family Conflict in India? – Gender or Gender Role Ideology. In Liisa Makela and Vesaa Sutari (eds), *Work and Personal Life Interface in the International Career Context*, Springer: Berlin, 71–93.

Ramadoss, K., and Rajadhyaksha, U. (2012). Gender Differences in Commitment to Roles, Work-Family Conflict and Social Support. *Journal of Social Sciences* 33(2): 227–233.

Ramesh, M. (2022). *'I'd Never Allow My Sister to Be Pilot': Women Pilots Face Everyday Sexism*. The Quint. 27 August 2022. Available at: https://www.thequint.com/gender/women/indian-women-pilots-fighting-everyday-sexism-misogyny#read-more. Accessed on September 7, 2024.

Rao, S.S. (2013). Structural Exclusion in Everyday Institutional Life. Labelling Of Stigmatized Groups in an IIT. In G.B. Nambissan and S.S. Rao (eds), *Sociology of Education in India. Changing Contours and Emerging Concerns*, Oxford Academic: Oxford, January 24, 2013, 199–223.

Rapoport, R., Bailyn, L., Fletcher, J., and Pruitt, B. (2002). *Beyond Work-Family Balance: Advancing Gender Equity and Workplace Performance*. Jossey-Bass: San Francisco, CA.

Rees, Teresa. (2004). Measuring Excellence in Scientific Research: The UK Research Assessment Exercise. In Delilah Al-Khudhairy, Nicole Dewandre, Helen Wallace (eds) *Gender and Excellence in the Making*. EUR21222. Brussels: DG-research, European Commission.

Ridgeway, C.L., and Correll, S.J. (2000). Limiting Inequality through Interaction. Utopian Visions: Engaged Sociologies for the 21st Century. *Contemporary Sociology* 29(1): 110–120.

Rosser, S. (2004). *The Science Glass Ceiling: Academic Women Scientists and the Struggle to Succeed*. Routledge: New York, NY.

Rudra, S. (2018). *Gender-Responsive Budgeting: A Task Ahead for India's 15th Finance Commission*. ORF Occasional paper, August 10, 2018. Available at: https://www.orfonline.org/research/gender-responsive-budgeting-a-task-ahead-for-india-s-15th-finance-commission. Accessed on September 7, 2024.

RVP (2025). Rashtriya Vigyan Puraskar 2025. Ministry of Home Affairs, GOI. Available at: https://awards.gov.in/Home/AwardLibrary. Last accessed on November 17, 2024.

S&T Indicators (2023). S&T Indicators Tables 2022-23. March 2023. Department of Science & Technology. Government of India, New Delhi. Available at https://dst.gov.in/sites/default/files/Updated%20ST%20INDICATORS%20TABLES%202022-23.pdf.Last accessed on November 16, 2024.

Saxena, R. (2022). *At 12.4 per cent, India Has Twice the Number of Female Pilots as the US*. Business Standard, August 9, 2022. Available at: https://www.business-standard.com/article/current-affairs/with-12-4-female-pilots-india-has-twice-as-many-women-pilots-than-us-122080900112_1.html. Accessed on September 7, 2024.

Schmidt, E.K., and Ovseiko, P.V. (2020). Acknowledging Complexity in Evaluation of Gender Equality Interventions. *EClinicalMedicine* 28(November 2020): 100623. Available at: https://www.thelancet.com/action/showPdf?pii=S2589-5370%2820%2930367-9. Accessed on September 7, 2024.

Schmidt, E.K., Ovseiko, P.V., Henderson, L.R., and Kiparoglou, V. (2020). Understanding the Athena SWAN Award Scheme for Gender Equality as A Complex Social Intervention in A Complex System: Analysis of Silver Award Action Plans in A Comparative European Perspective. *Health Research Policy and Systems* 18(19) 1–21. Available at: https://doi.org/10.1186/s12961-020-0527-x. Accessed on September 7, 2024.

Shah, A. (2018). India Soars Above Global Average in Hiring Female Airline Pilots. *Reuters*, September 5, 2018. Available at: https://www.reuters.com/article/uk-india-pilots-women-analysis-idUKKCN1LL0PF. Accessed on September 7, 2024.

Shen, H. (2013). Inequality Quantified: Mind the Gender Gap. *Nature* (News Feature), March 6, 2013. Available at: https://www.nature.com/news/inequality-quantified-mind-the-gender-gap-1.12550. Accessed on September 7, 2024.

Singh, N. (2018). Men Too Hit by Guilt of Work-Life Balance. *The Times of India*, September 13, 2018. Available at: https://timesofindia.indiatimes.com/business/india-business/men-too-hit-by-guilt-of-work-life-balance/articleshow/65789765.cms. Accessed on September 7, 2024.

Sinha, A. (2023). Science Congress: Why Its Glory Days Are Over. *The Indian Express*, January 5, 2023. Available at: https://indianexpress.com/article/explained/explained-sci-tech/indian-science-congress-nagpur-narendra-modi-8359644/. Accessed on September 7, 2024.

Sinha, U.B., and Sinha, D. (2011). Are Indian Women Scientists Victims of the 'Glass Ceiling'? *Current Science* 100(6): 837–840.

Srivastava, S. (2018). Masculinity Studies and Feminism: Othering the Self, Engaging Theory. In R. Choudhary and Z.A. Baset (eds), *Men and Feminism in India*, 1st edition (Hardback), Routledge: London, 35–49.

Srivastava, A. (2023). JEE Main 2023 Topper: 'Boys Said You Will Get IIT Due to Supernumerary Seats, I Aimed and Achieved 100 Percentile'. *The Indian Express*, May 3, 2023. Available at: https://indianexpress.com/article/education/jee-main-2023-topper-boys-said-you-will-get-an-iit-seat-under-quota-i-achieved-100-percentile-jeemain-nta-nic-in-jeeadv-ac-in-8584014/. Accessed on September 7, 2024.

STIP (2020). *Science, Technology and Innovation Policy.* Draft STIP Doc 1.4, Department of Science and Technology, Government of India, December 2020. https://dst.gov.in/sites/default/files/STIP_Doc_1.4_Dec2020.pdf. Accessed on September 7, 2024.

Subrahmanyan, L. (1998). *Women Scientists in the Third World: The Indian Experience.* Sage Publications: New Delhi.

Sur, Abha. (2009). Dispersed Radiance. Women Scientists in C.V. Raman's Laboratory. In Neelam Kumar (ed), *Women and Science in India,* OUP: New Delhi, 98–136.

Tellis, A. (2000). Resisting Feminism. Indian Journal of Gender Studies 8(1): 117–119.

Tewari, R., and M. Bhardwaj, M. (2018). *Mapping Patents and Research Publications of Higher Education Institutes and National R&D Laboratories of India.* DST-Centre for Policy Research, Panjab University, Chandigarh. At http://www.dst.gov.in/sites/default/files/FULL%20BOOK-Chandigarh.pdf. Accessed on September 7, 2024.

Thomas, R. (2020). Brahmins as Scientists and Science as Brahmins' Calling, Caste in An Indian Scientific Research Institute. *Public Understanding of Science* 29(3): 306–318.

UN (2002). *Gender Mainstreaming. An Overview.* United Nations: New York. Available at: https://www.un.org/womenwatch/osagi/pdf/e65237.pdf. Accessed on September 7, 2024.

UNESCO. (2017). *Measuring Gender Equality in Science and Engineering: The SAGA Toolkit*. Working Paper 2. UNESCO. Available at: https://uis.unesco.org/sites/default/files/documents/saga-toolkit-wp2-2017-en.pdf. Accessed on September 7, 2024.

UNESCO (2021). *Transforming MENtalities: Gender Equality and Masculinities in India*. By Coley, Christopher, Gressel, Christie M. and Bhavani, Rao. UNESCO: New Delhi. Available at: https://unesdoc.unesco.org/ark:/48223/pf0000377859.locale=en. Accessed on September 7, 2024.

UNESCO (2022). *A Braided River: The Universe of Indian Women in Science*. By C. Coley, A. Dhillon, C. Gressel. UNESCO: New Delhi.

UNESCO Science Report (2021). *To Be Smart, The Digital Revolution Will Need to Be Inclusive: Excerpt from The UNESCO Science Report*. By Bello, Alessandro, Blowers, Tonya, Schneegans, Susan, Straza, Tiffany. UNESCO: Paris. Available at: https://unesdoc.unesco.org/ark:/48223/pf0000375429. Accessed on September 7, 2024.

Vaid, M. (2019). *ISRO's Magnificent Women and Their Flying Machines*. Speaking Tiger Books LLP: New Delhi.

Vaidya, Vidita A. (2017). Towards a Narrative of Gender in the Biological Sciences. *Economic and Political Weekly* 52(7): 83.

van den Brink, M., Benschop, Y., and Jansen, W. (2010). Transparency in Academic Recruitment: A Problematic Tool for Gender Equality? *Organization Studies* 31(11): 1459–1483.

Wallon, G., Bendiscioli, S., and Garfinkel, M.S. (2015). *Exploring Quotas in Academia*. Robert Bosch Stiftung project number: 12.5.8050.0239.0. EMBO: Heidelberg. Available at: https://www.embo.org/documents/science_policy/exploring_quotas.pdf. Accessed on September 7, 2024.

WFP (2020). *Evaluation of the Gender Policy (2015–2020). World Food Programme*. May 23, 2020. Available at: https://www.wfp.org/publications/evaluation-gender-policy-2015-2020. Accessed on September 7, 2024.

Winchester, H., S. Lorenzo, L. Browning, and C. Chesterman. (2006). Academic Women's Promotions in Australian Universities. *Employee Relations* 28 (6): 505–522.

Yang, T.Y.T. Yang, Woodruff, Teresa K., and Uzzi, Brian. (2022). Gender-Diverse Teams Produce More Novel and Higher-Impact Scientific Ideas. *PNAS* 119(36) e2200841119. Available at: https://www.pnas.org/doi/10.1073/pnas.2200841119. Accessed on September 7, 2024.

INDEX